NASSAU 2006

NASSAU 2006

Proceedings of the 4th Nassau Mössbauer Symposium held in Garden City, New York, USA, 13–14 January 2006

Edited by

E. E. ALP

Argonne National Lab, Argonne, IL, USA

and

C. I. WYNTER

Nassau Community College, Garden City, NY, USA

Reprinted from *Hyperfine Interactions*
Volume 170, Nos. 1–3 (2006)

Springer

A C.I.P. Catalogue record for this book is available from the Library of Congress.

978-3-540-71124-7

Published by Springer
P.O. Box 990, 3300 AZ Dordrecht, The Netherlands

Sold and distributed in North, Central and South America
by Springer
101 Philip Drive, Norwell, MA 02061, U.S.A.

In all other countries, sold and distributed
by Springer
P.O. Box 990, 3300 AZ Dordrecht, The Netherlands

Printed on acid-free paper

Table of Contents

Hyperfine Interact (2006) 170:1–3
DOI 10.1007/s10751-006-9478-z

PREFACE

4th Nassau Mössbauer symposium analysis and summary

E. Ercan Alp · Clive I. Wynter

Published online: 11 January 2007

There was warm, brilliant sunshine on January 13th in Garden City, New York, which was a good omen for what the Symposium would be like, and the weather indicator was true to prediction.

After a light breakfast, a welcome speech was delivered to participants by Academic 13 Vice President of the Nassau Community College, Dr John Ostling, Co-Chair of the Conference. Dr. Ercan Alp introduced the first speaker, Professor Moshe Pasternak (Tel Aviv, Israel), who delivered his authoritative lecture on high pressure Mössbauer experiments, which breaks new ground into physics of correlated electron systems. This was followed by an exciting talk by Professor Elisa Baggio-Saitovitch (CBPF, Brasil) in which she has described magnetism in RNi_2B_2C system. Professor Werner Keune (Duisburg, Germany) ended the first session with an articulate discussion of lattice dynamics of amorphous materials, $Fe_{1-x}Al_x$, $Fe_{1-x}Mg_x$, $Fe_{1-x}Sc_x$, $Fe_{1-x}Tb_x$, $Fe_{1-x}Y_x$, $Fe_{1-x}Sn_x$, $Fe_{1-x}Si_x$ and $Fe_{1-x}Dy_x$ systems, as measured by the phonon density of states using synchrotron radiation.

The second session of the day was chaired by Dr. Clive Wynter and Professors Timothy Sage, Stephen Cramer and Robert Scheidt "shed new light" on proteins and iron (II) porphyrinates as model hemoglobin-type compounds. While Profs Sage (Northeastern University) and Cramer (University of California at Davis) focused on vibrational spectroscopy of proteins and enzymes with synchrotron radiation based experiments, Prof Scheidt (Notre Dame) presented their work on Mössbauer spectroscopy of imidazole and imidazolate porphyrins.

Dr Gopal Shenoy, revered for his contribution to Mössbauer Spectroscopy, with Dr Stan Ruby, for whom the conference was dedicated, chaired the next session and introduced our keynote speaker, Dr Goestar Klingelhofer (Mainz, Germany). Having accomplished a life-long dream of not one but two Mössbauer spectrometers on Mars, he gave us insight on how the experiments are planned, conducted, and the data are evaluated. The analysis of the

E. E. Alp (✉)
Argonne National Lab, Argonne, IL, USA
e-mail: alp@anl.gov

C. I. Wynter
Nassau Community College, Garden City, NY, USA

data obtained on Mars is still continuing and the results and contribution to date to Geochemistry has injected "new life" into "table top" Mössbauer Spectroscopy. It is clear that the exciting results obtained so far will keep Mössbauer spectroscopy on the front page for the foreseeable future, particularly because the data obtained, combined with a large data base on minerals obtained through 40 years of Mössbauer spectroscopy will surely provide some unique answers to pre-biotic and biotic activity on Mars, in the coming months. Dr Klingelhofer also mentioned that the converted data, suitable for analysis is available through their web site. As of the conference date on January 13, 2006, there were more than 250 sets of data obtained from Rover called Spirit which landed at 180° longitude, and 210 sets from the Rover called Opportunity, which landed on the opposite side at 0 longitude. The unmistakable signal of presence of water through detecting the presence of Jarosite was the highlight of the last years data, while detecting a meteorite was considered rather a lucky coincidence.

Dr Wolfgang Sturhahn had the unenviable task of following Dr Klingelhofer, but he gave a mastery lecture on the perspectives on nuclear resonant spectroscopy using synchrotron radiation, generating enough excitement for recent accomplishments of measurement of speed of sound at 1,500°C and 1.5 Mbar pressure range.

The last session on the first day on Friday was chaired by Professor John Stevens. Dr Ravi Kukkadapu (Pacific Northwest Lab) spoke eloquently on Mössbauer spectroscopy in clay mineral research regarding bioreduction of Fe with bacteria, this was followed by an instructive lecture by Professor Dominic Ryan (McGill, Canada) on studying surfaces interfaces and thin films by Conversion Electron Mössbauer Spectroscopy, and some applications in dual phase steels that harden as they are deformed. Professor Georges Denes (Concordia University, Canada) was impressive in his application of the VSEPR theory to tin (II) stereochemistry and order–disorder transition in organometallic compounds, which would be difficult or impossible to detect using X-ray or neutron diffraction.

A delightful, tasty dinner followed at the Westbury Manor; Dr Stan Ruby's son, Walter Ruby and his fiancée were our guests of honor. Walter was able to give us great "insight" into his father's enthusiasm and commitment to Mössbauer Spectroscopy.

Although the weather was not as kind on Saturday, the momentum of the Friday sessions continued to prevail.

The first session was chaired by Dr Sturhahn, who introduced Dr Shenoy; a colleague and close friend of Stan Ruby. Dr Shenoy (Argonne National Laboratory) explained the wonderful contribution of "Stan" as he was affectionately called, in terms of his early accomplishments developing sources for ^{40}K, ^{83}Kr, ^{121}Sb, ^{238}U, ^{236}U, and ^{243}Am. Stan also was busy investigating the affect of ultrasound excitations on nuclear decay, and through such studies he came up with an idea that became almost synonymous with his name: nuclear resonance scattering using synchrotron radiation. Dr Shenoy provided some insight into the early days of this rather impossible task. It was truly a reflection of the life and legacy of this great nuclear physicist. Dr Jennifer Jackson (Carnegie Institute of Washington) gave an inspiring talk on novel melting investigations of iron at high-pressure using synchrotron Mössbauer Spectroscopy and Dr Jiyong Zhao (Argonne National Laboratory) continued to elucidate the phonon density of states of ^{83}Kr under high pressure using NRIXS.

Professor Brent Fultz chaired a lively session starting with Professor Roberto Mercader (La Plata, Argentina) discussing nanofabrication techniques and elaborated on the magnetic and hyperfine properties of these microscopic particles, using template assisted electrodeposition technique. Prof Mercader also gave an interesting account of early history of physics education in Argentina at the dawn of the twentieth century. Dr Michael Hu (Carnegie Institute of Washington) examined $BiFeO_3$ under high pressure using nuclear forward scattering and X-ray emission spectroscopy. Dr Kailash Mishra (Osram-Sylvania) explained enthusiastically the phosphor properties of Eu^{2+} and Eu^{3+} and the degradation processes associated with this phosphor using ^{151}Eu Mössbauer spectroscopy. Professor Eugene Hoffman (Morgan State University) discussed the Mössbauer spectroscopy of high-Ca pyroxene specimens which showed some anomalies in the near-infrared spectra.

The final session was chaired by Professor Werner Keune. It started with a lively and authoritative interpretation of Mössbauer Diffractometry by Professor Brent Fultz (California Institute of Technology) as it is applied to order–disorder in Fe–Al system; this was followed by B. Roldan-Cuenya (University of Central Florida) who used nuclear resonant scattering to show dramatic differences between the phono-DOS of bulk iron and that of size-selective nanoparticles. Professor David Agresti (University of Alabama at Birmingham) discussed velocity calibration for Mars Mössbauer data while Professor Zameer Hasan (Temple University) was enlightening in his interpretation of nuclear forward scattering vs. transmission Mössbauer spectroscopy studies of atomically tailored Eu-based materials. Professor John Stevens (University of North Carolina at Asheville) closed the symposium on a note of optimism and caution. While the meeting was clearly successful in its high quality, innovation and enthusiasm, he cautioned that we must continue to recruit more graduate students and scientists in the field. Prof Stevens mentioned that there are 100 "Mössbauer" papers/year from the United States, which represented 12% of the world publications. He also mentioned that the number of researchers that have published at least 5 papers in the last 5 years were 141, and 28 researchers published 10 papers or more in the last 5 years. He has proposed scheduling regular conferences, a Newsletter, web-based virtual conferences, and a web-site for communication as alternatives to increase the communication between Mössbauer researchers.

Twelve posters were viewed from lunch time Friday to the end of the Conference. A wide variety of topics were covered including Professor George Eng's use of Mössbauer to determine the effect of salinity on the speciation of several triorganotin compounds. Professor Georges Denes study of $BaSnF_4$, Laura Perry studies *of anistropic* contribution to the transferred hyperfine field using a field-induced spin-reorientation; Yisong Guo demonstrated the use of synchrotron radiation based (NRS) of iron and nickel compounds in biological systems. Dr S. Kharlamova used high-pressure x-ray, optical studies and ^{57}Fe synchrotron Mössbauer spectroscopy measurements to examine single crystals of $GdFe_3(BO_3)_4$.

Professor Anthony Kinyua investigated *Phragmites australis* and *Spartina alaterniflora* while Dr Roshan Lal used ^{57}Fe Mössbauer spectroscopy to study the effect of zinc oxide substitution in glass system. Professor Clive Wynter displayed the use of ^{151}Eu Mössbauer spectroscopy in the study of europium salen complex nanoparticles. Chris Voyer displayed a complete solution to the Mössbauer problem, all in one place.

This meeting was truly a fest for first, second and third generation Mössbauer spectroscopists and scientists interested in new learning of quantum phenomena. The low-key and serene atmosphere proved to be an effective medium for better interaction between established researchers and newcomers, which was clearly a refreshing break from large meetings. The participants expressed their appreciation of the excellent facilities and services provided by Nassau Community College and their friendly staff.

Hyperfine Interact (2006) 170:5–13
DOI 10.1007/s10751-006-9477-0

Scientific legacy of Stanley Ruby

G. K. Shenoy

Published online: 9 January 2007

Abstract Stanley L. Ruby (1924–2004) made major contributions to Mössbauer spectroscopy and was the first to suggest the feasibility of observing the Mössbauer effect using synchrotron radiation. In this article we recall his scientific legacy that have inspired his scientific colleagues.

Key words annihilation of acoustic phonons · isoelectronic compounds · synchrotron radiation · Mössbauer effect

1 Introduction

During his active scientific career, Stanley Ruby always addressed most fundamental aspects of physics without being a "reductionist" – asking questions regarding the fundamental interaction of electromagnetic waves with matter – nucleons, atoms, molecules and the condensed matter, and answering these questions with unique demonstration experiments. This *recollection* paper is intended to be a brief compilation of Stanley Ruby's selected contributions. They are creation and annihilation of acoustic phonons, chopped gamma-rays, and jump diffusion in condensed matter. In addition, Ruby's nuclear physics background fitted best for him to identify many new candidate nuclear resonant transitions across the nuclear periodic table; and his grasp of chemical physics led to the development of isomer shift systematic among isoelectronic compounds of 5s–5p atoms. Finally, Ruby's most pragmatic suggestion in 1974 to use synchrotron radiation to observe nuclear resonance, and its eventual demonstration, addressed many fundamental aspects of collective nuclear excitation process and gamma-ray coherence. Further more it opened many new avenues of application of the tool to a broader area of science which otherwise would have been impossible with the traditional approach of using radioactive sources in performing Mossbauer spectroscopy measurements.

G. K. Shenoy (✉)
Argonne National Laboratory, Argonne, IL 60439, USA
e-mail: gks@aps.anl.gov

2 Creation and annihilation of acoustic phonons

Soon after the discovery of the Mossbauer effect it became clear that nuclear resonant absorption measurement could provide valuable information on the lattice dynamics. One considered the Mossbauer effect to be a potential tool for the investigation of phonon spectra, in many ways similar to slow neutron scattering or Rayleigh scattering of light or scattering of conduction electrons by phonons. To obtain an insight into this, we use the well known formalism of Singwi and Sjolander [1] where the Mossbauer effect is completely described by the Fourier transform of the 'intermediate scattering function,' $F_s(\kappa, t)$. The Mossbauer resonance cross-section is given by

$$\sigma_a(E) = \frac{\sigma_0 \Gamma}{4h} \int \exp\left[i\omega t - (\Gamma/2h)|t|\right] \times F_s(\kappa, t) dt \tag{1}$$

where $h\omega = E - E_0$ and κ is the gamma-ray momentum vector. The intermediate scattering factor is defined as

$$F_s(\kappa, t) = \frac{1}{N} \sum \left\langle e^{-i\vec{\kappa}\cdot\vec{r_i}(0)} e^{-i\vec{\kappa}\cdot\vec{r_i}(t)} \right\rangle_T \tag{2}$$

Substituting the value of $F_s(\kappa, t)$ for an isotropic harmonic solid [2], we obtain

$$\sigma_a(E) = \frac{\sigma_0 \Gamma}{4h} \exp(-2W) \int \exp\left[i\omega t - (\Gamma/2h)|t|\right][1 + f(\omega)] dt \tag{3}$$

where $\exp(-2W)$ is the normal Debye–Waller (or Lamb–Mossbauer) factor and $f(\omega)$ describes the higher terms representing the creation and annihilation of phonons. The first term (1 in second square bracket) in Eq. 3 represents the zero-phonon peak (the Mossbauer effect) and it has the largest contribution to the cross section over a very narrow energy band determined by the width of the nuclear resonance (of few neV for ^{57}Fe resonance). The second term $f(\omega)$ gives the frequency distribution of the phonon normal modes. This contribution to the cross-section is feeble and spread over a broad energy range with a cut-off energy determined by the Debye temperature (typically a few meV). Hence it is difficult to measure the second term (with a typical Mossbauer resonance line width of a few neV) although it has important dynamical information of the system.

In the *Physical Review Letters* published in 1960, Ruby and Bolef [3] point out that ' ... *a direct attempt at such study* (off resonance measurement of phonon spectrum) *seems difficult since it requires the measurement of nuclear γ-ray absorption cross sections much smaller than the photoelectric cross section for the same atom. In an attempt to investigate the interactions between phonons and emitting nuclei, therefore, it was decided to generate low-energy phonons acoustically, and to study their effect on γ-ray spectrum.*' The set up used to create low frequency acoustic phonons is shown in Fig. 1a, where the frequency of the resonant gamma-rays was phase modulated by a piezoelectric transducer. This effectively produced 'off resonance' absorption peaks at energies of various harmonics of the modulation frequency (Fig. 1b). In effect, this would be equivalent to an Einstein solid had very low-energy acoustic phonons.

The measured resonance peaks observed in this experiment can be understood by describing the displacement of the vibrating source with frequency Ω by

$$r_i(t) = r_{\max} \sin \Omega t \tag{4}$$

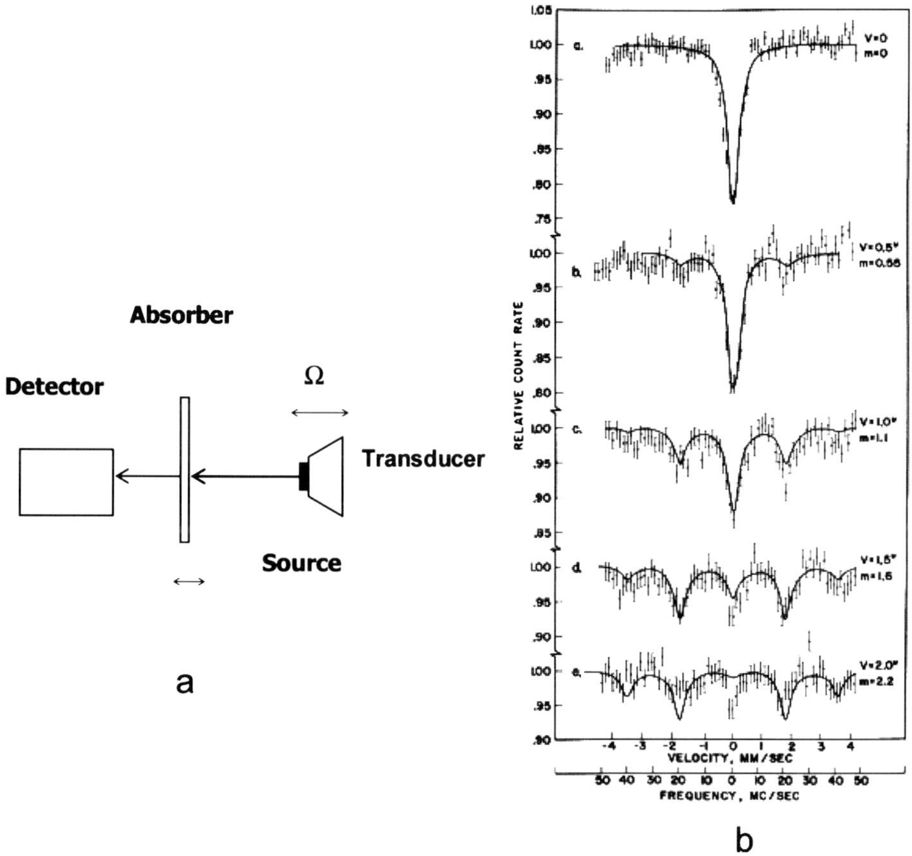

Fig. 1 **a** Experimental setup in which the ^{57}Co source is mounted on a piezoelectric transducer vibrated at frequency Ω. A single-line stainless steel absorber is Doppler shifted to measure the spectrum. **b** The spectra measured for different values of m for a given transducer frequency of 20 MHz [3]

The resulting expression for Mossbauer resonance cross section (Eq. 1) is

$$\sigma_a(E) \approx \frac{\sigma_0 \Gamma}{4h} \sum_n J_n^2(\kappa \cdot r_{\max}) \frac{\Gamma}{(E - E_0 \pm n\Omega)^2 + \Gamma^2/4} \quad n = 0,\ 1,\ 2,\ 3\ldots \tag{5}$$

The value of $n=0$ gives the central zero-phonon peak, while the other integral values of n produces the side bands in the off resonance part of the energy spectrum. The amplitude of the Lorentzians is proportional to the second power of the Bessel function, $J_n(\kappa \cdot r_{\max})$. The acoustic phonon peaks are located at $\pm n\Omega$ where the transducer frequency Ω=20 MHz (~83 neV or 1.7 mm/s). The intensity of the absorption peaks is governed by the argument of the Bessel function, which is defined by $m = \kappa \cdot r_{\max} = \left(^{v_m}/_c\right)\left(^{\Omega}/_{\omega_0}\right)$, where v_m is the velocity of the transducer. In Fig. 1b, the intensity of the central (n=0) resonance peak decreases as the value of the m is increased by changing the transducer velocity. In fact the term $\sum_n J_0^2(\kappa \cdot r_{\max})$ is equal to the Debye–Waller factor $\exp\left(-\langle r^2 \rangle \kappa^2\right)$ [4]. The intensity of the acoustic phonon peaks (n=1, 2. 3...) increases with increasing m.

In every measure, this is a classic demonstration that answered many fundamental questions regarding the interaction of the resonant emission of gamma-rays by nuclei bound in a crystal and creation of phonons. There is also an interesting proposal to apply this technique to practical problems involving the interaction of acoustic waves with inhomogeneous systems of frozen globular proteins to investigate their elastic properties [5]. It is of interest to note that it took another two decades before which one could routinely probe the phonon spectrum using nuclear resonant scattering with synchrotron radiation (see Section 6).

3 Chopped gamma rays

It was recognized early on that if the resonant gamma-ray travels through a medium with a different index of refraction, it will change the phase of the wave, and if this phase changes with time, it will shift the frequency of the resonant gamma-ray [6]. In order to demonstrate this Ruby et al. [7] introduced a composite of thickness L containing plastic and copper between the source and absorber. As can be seen from Fig. 2a, the time varying path length through copper is given by $x(t,T_o)$ and that through plastic is given by $(L\text{–}x(t,T_o))$. The function $x(t,T_o)$ can be precisely determined from the geometrical details of the composite and the rate of variation of $x(t,T_o)$ perpendicular the gamma-ray direction. Using Eq. 2, the phase and amplitude variation can be expressed in terms of intermediate scattering function,

$$F_s(\kappa,t)=\int\left(e^{-i\kappa\,X(0,T_0)}e^{-i\left(\kappa_c-\kappa_p\right)X(t,T_0)}\right)dX \tag{6}$$

where κ_c and κ_p are the gamma-ray momentum vectors in copper and in plastic. The spectrum for a given time variation of $x(t,T_o)$ can be calculated with the using the Eq. 1. In Fig. 2b, the spectra measured at various values of the angular velocity, ω, of the composite wheel is shown. At a given value of ω, both the amplitude and the phase of the resonant gamma-rays vary as it travels through the composite. The modulation frequency is given by $2\pi\omega r/s$ where r is the radius of the wheel and s is the separation between the copper wires. Ruby et al. [7] have presented the full analysis of the measured spectra shown by the solid lines through the measured data.

This powerful capability of Mossbauer spectroscopy allows one to investigate the dynamics in systems which do not contain Mossbauer resonating nuclei. The study of the variation in both the phase and amplitude contrast, for example in a biological system, will be useful in understanding the dynamical correlations. Regrettably, this potential has not been exploited since the pioneering works of Grodzins et al. [6] and Ruby et al. [7].

4 Diffusion of ions in cooled liquids

The theoretical work of Singwi and Sjolander [1] inspired early experimental work on the diffusion of Mossbauer nuclei in liquids. Ruby et al. [7] have investigated the diffusion process in liquids by comparing the microscopic (^{57}Fe Mossbauer) and macroscopic diffusion measurements. Their interest was to probe whether the ions diffuse in liquids by taking large jumps (so called jump diffusion model) or through small frequent jumps (so called continuous diffusion model). We refer to a classic work by Singwi and Sjolander [8] which addresses theoretical study of diffusive motion using neutron scattering, but is equally applicable to Mossbauer investigations. In this work the diffusive motion is considered to be made up of consecutive steps in which the atom firstly occupies position A

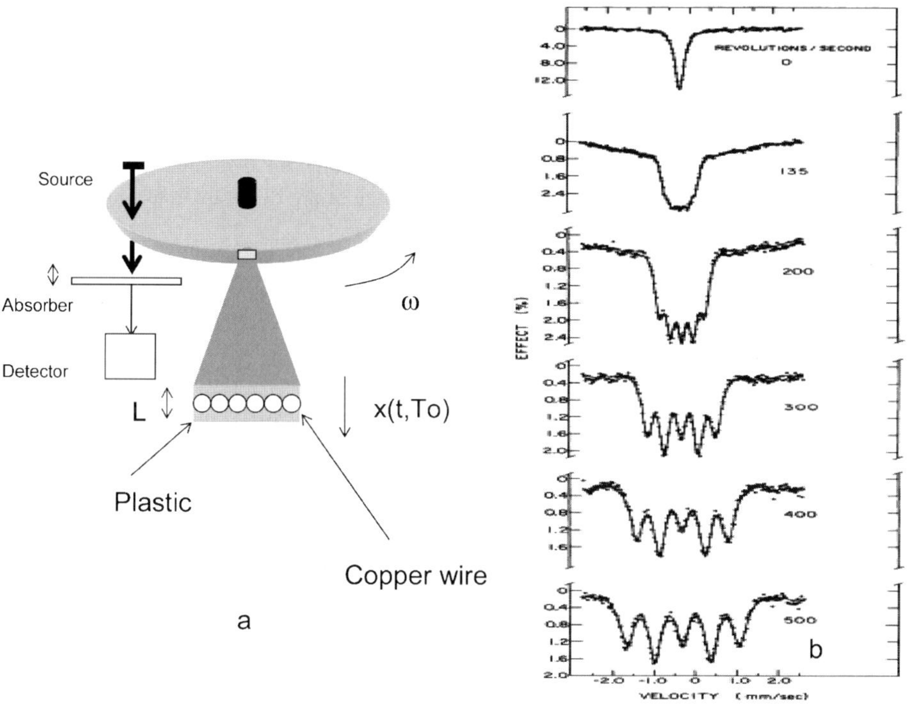

Fig. 2 **a** Experimental setup for varying the phase and amplitude modulation realized by rotating the composite wheel made up of plastic and copper. **b** The measured spectra measured with various angular velocity of the composite wheel. The *solid line* describes the phase and amplitude modulation [7]

where it performs an oscillatory motion followed by diffusion to position B where it again performs oscillatory motion, and so on. The two characteristic times that are involved in this diffusion model are τ_0 characterizing the mean dwell time of the atom at locale A before it jumps to position B, and τ_1 characterizing the diffusion time from position A to B separated by a mean distance r_0. It can be shown that the Mossbauer resonance line shape from this diffusive motion is given by a broadened Lorentzian:

$$\sigma_a(E) \propto \frac{\sigma_0 \Gamma}{4h} e^{-2W} \frac{\Gamma}{(E-E_0)^2 + (\Gamma + \Delta E)^2 / 4} \tag{7}$$

where ΔE is the diffusion line width. When the diffusive time is very short compared to the mean dwell time time, $\tau_1 \ll \tau_0$, the value of ΔE is given by

$$\Delta E = \frac{2\hbar}{\tau_0}\left[1 - \frac{e^{-2W}}{1 + \kappa^2 D \tau_0}\right] \tag{8}$$

If the jump lengths are small compared to the resolving power of the Mossbauer resonant radiation, $r_0 \ll 1/\kappa$ (i.e., $\kappa^2 D\tau_0 \ll 1$), $\Delta E \approx 2\hbar\kappa^2 D$. This is the familiar case of Brownian motion with an intermediate scattering function given by $F_s(\kappa, t) = e^{-\kappa^2 D|t|}$.

On the other hand if the mean diffusion length is large, $r_0 \gg 1/\kappa$ (i.e., $\kappa^2 D\tau_0 \gg 1$), the increased line width is given by $\Delta E \approx 2\hbar\tau_0$. With this background we can address some of the experimental results.

In the early work of diffusion of Sn metal near the melting point [9], the diffusion line width data was analyzed using the continuous diffusion model. On the other hand, the diffusion of iron in glycerin [10] was analyzed using a jump diffusion model. Ruby et al. [7] hence addressed the nature of the diffusion in a liquid by comparing both macroscopic and microscopic diffusion measurements. They univocally confirmed the applicability of the continuous diffusion model and also demonstrated for the first time the evidence of Arrhenius-like behavior of the temperature dependence of the diffusion constant in a liquid over six decades.

In spite of the advance made by Ruby et al. in using the Mossbauer effect to understand diffusion in liquids, the field has been some what stagnant. Recent use of synchrotron radiation perturbed angular correlation (see for example [11]) has revived this field and provided a broad scope for the future.

5 Unique Mossbauer resonance studies

When there is no radioactive parent to populate the resonant state, the observation of the Mossbauer effect requires alternative approaches. These include Coulomb excitation or nuclear reaction between the nucleus and an accelerated particle or through neutron capture by the nucleus at a reactor. Ruby and Holland [12] were the first to demonstrate this approach by performing a ^{39}K(d,p)^{40m}K reaction to populate the 29.8-keV first-excited state in ^{40}K. The nuclear level scheme is shown if Fig. 3.

The consequences of dislodging the resonant nuclei from the lattice location following a nuclear reaction were first discussed in this paper. The measured resonance with a potassium metal source and a potassium chloride absorber is also shown in the Fig. 3. It should be pointed out that this investigation was further challenging from the low value for the natural abundance of ^{40}K.

The 29.8 keV resonance in ^{40}K was also performed by populating the state by ^{39}K $(n,\gamma)^{40m}$K reaction [13]. In this reactor based studies the emphasis was the determination of isomer shift for various potassium compounds to deduce the change the nuclear charge radius on excitation.

Over the next few years Ruby in collaboration with many scientists investigated new and exotic Mossbauer resonances. The focus of many of these studies was to establish the properties of nuclear levels involved in the Mossbauer transitions. These include 9.3 keV transition in ^{83}Kr [14, 15], transitions in isotopes of 5s–5p elements (^{119}Sn, ^{121}Sb, ^{125}Te, ^{127}I, ^{129}I, ^{129}Xe) [16, 17], 44.7 keV transition in ^{238}U [18], and 83.9 keV transition in ^{243}Am [19].

6 Mossbauer spectroscopy using synchrotron radiation

A remarkable piece of research was presented by Ruby [20] during the first week of September 1974 at the *International Conference on the Applications of the Mossbauer Effect* held in Bendor, France, in which it was pointed out how nuclear coherent scattering can be observed using a synchrotron radiation source. The major challenge was to separate the resonant Mossbauer radiation with narrow energy band width (~10^{-9} eV) from the background radiation from monochromatized synchrotron radiation with a band width of about eV. The essence of Ruby's suggestion is to obtain this discrimination in the time

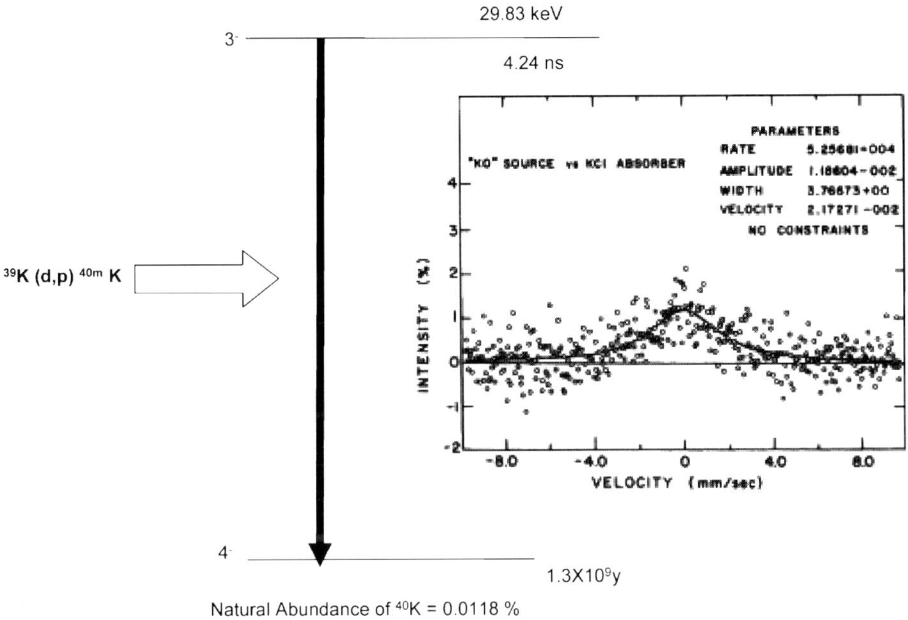

Fig. 3 Nuclear level scheme of ^{40m}K and the observed 29.8 keV Mossbauer resonance spectrum reported in [12]

domain rather than in the energy domain by using the time structure of the synchrotron radiation source, taking advantage of long life time of the nuclear excited (Mossbauer) level compared to those of other electronic scattering processes.

This remarkable suggestion has flourished into a diverse active research field during the past 30 years with new avenues that were not available with the use of conventional radioactive sources to perform Mossbauer spectroscopy measurements. Nearly 10 years after this suggestion, numerous groups made progress in the field. It became clear from the experiments performed at PEP storage ring at Stanford that an undulator source is best suited to carry out nuclear resonant scattering studies [21, 22]. The third generation synchrotron radiation sources designed and built soon after (such as APS at Argonne, ESRF at Grenoble, and SPring-8 at Harima City) incorporated beamlines based on undulator sources to perform such experiments. Interested reader should refer to [23] which has a host of very valuable articles.

Perhaps somewhat less recognized but most important contribution to the success of nuclear coherent scattering is Ruby's suggestion on the use of avalanche photon detectors (APDs) which provided the required time resolution [24–27]. The APDs have become the important tool in many areas of synchrotron radiation scattering studies.

In the coherent nuclear scattering experiment using synchrotron radiation one measures the intermediate scattering function, $F_s(\kappa, t)$, where the value of κ is the momentum vector of the resonant gamma-ray. There are subtle but very important differences in the details of this function in the context of nuclear coherent scattering, since the intermediate scattering function now represents the collective excited state of an ensemble of nuclei rather than that of a single nucleus. A spatially coherent superposition of excited states from such an ensemble forms a *nuclear exciton*. If there are magnetic and/or quadrupole hyperfine fields present at these nuclei, the time development of the re-emitted wave packet from nuclei will produce a complex beat pattern due to interference between coherently excited hyperfine levels [21, 22]. Indeed the analysis of hyperfine interactions measured in this fashion using the time domain spectroscopy became far more convoluted than that obtained from the energy spectroscopy performed using conventional Mossbauer spectroscopy. On the other hand methodologies to analyze the time spectrum in so called forward scattering geometry are now been well developed [28]. In the subsequent works, Ruby played a central role to demonstrate the unique capabilities of this new tool [29, 30].

It is worth noting that the investigation of interaction of acoustic waves with resonant gamma-ray performed by Ruby and Bolef [3] using conventional radioactive source (see Section 2), has been repeated in the forward scattering geometry using synchrotron radiation source [31]. In the new approach one investigates the interaction of the acoustic waves with nuclear exciton. The anticipated beat pattern in the time spectrum will not be seen in this case since the interference exactly cancels out when all the frequencies are present [32]. Absorption of a component of the energy spectrum by introducing a second absorber in the path of the resonant radiation will exhibit time spectrum due to quantum beats [31, 32].

During early 1990s, Stanley Ruby suggested that the phonon spectrum should become measurable by collecting incoherent photons from the nuclear deexcitation in the energy range spanning the phonon energies of a few tens of meV away from the resonance ($f(\omega)$in Eq. 3). This proposal got tested in 1995 by Seto et al. [33] and Sturhahn et al. [34] where the authors measured the phonon spectrum by scanning the photon energies on either side of the resonance over about 100 meV range with a resolution of few meV. These remarkable events opened a new area of nuclear inelastic resonance spectroscopy, and in recent years, it has been the main stay of this field [35, 36].

7 Conclusions

In this article, we have attempted to provide a summary of major scientific contributions by Stanley Ruby, and his impact to the field of Mossbauer resonance spectroscopy and nuclear resonance scattering. He has left a long term scientific legacy which has advanced science in many areas over many decades. This brief *recollection* of Stanley Ruby's scientific contributions is neither intended to be a complete review of the subject nor a scientific chronology of research contributions to this ever growing field. We apologize to those authors whose works might not have received due recognition in this article.

Acknowledgements This work was supported by the U.S. Department of Energy, Basic Energy Sciences, Office of Science, under Contract No. DE-AC02-06CH11357.

References

1. Singwi, K.S., Sjolander, A.: Phys. Rev. **120**, 2211 (1960)
2. Van Hove, L.: Phys. Rev. **95**, 249 (1954)
3. Ruby, S.L., Bolef, D.I.: Phys. Rev. Lett. **5**, 5 (1960) (Reprinted Figure 1 with permission from the authors. Copyright 1960 by the American Physical Society)

4. Frauenfelder, H.: The Mossbuer Effect. W. A. Benjamin, New York, p. 19, (1962)
5. Zatovsky, A.V., Zvelindovsky, A.V.: J. Phys. D: Appl. Phys. **27**, 839 (1994)
6. Grodzin, L., Phillips, E.A.: Phys. Rev. **124**, 774 (1961)
7. Ruby, S.L., Preston, R.S., Skov, C.V., Zabransky, B.J.: Phys. Rev. A **8**, 59 (1973) (Reprinted Figures 1 and 2 with permission from the authors. Copyright 1973 by the American Physical Society)
8. Singwi, K.S., Sjolander, A.: Phys. Rev. **119**, 863 (1960)
9. Boyle, A.J.F., Bunbury, D.St.P., Edwards, C., Hall, H.E.: Proc. Phys. Soc. **77**, 129 (1961)
10. Bunbury, D.St.P., Elliott, J.A., Hall, H.E., Williams, J.M.W.: Phys. Lett. **6**, 34 (1963)
11. Sergueev, I., van Bürck, U., Chumakov, A.I., Asthalter, T., Smirnov, G.V., Franz, H., Rüffer, R., Petry, W.: Phys. Rev. B **73**, 024203 (2006) (See for example)
12. Ruby, S.L., Holland, R.E.: Phys. Rev. Lett. **14**, 591 (1965) (Reprinted Figures 1 and 2 with permission from the authors. Copyright 1965 by the American Physical Society)
13. Tseng, P.K., Ruby, S.L., Vincent, D.H.: Phys. Rev. **172**, 249 (1968)
14. Hazony, Y., Hillman, P., Pasternak, M., Ruby, S.L.: Phys. Lett. **2**, 337 (1962)
15. Ruby, S.L., Selig, H.: Phys. Rev. **147**, 348 (1966)
16. Ruby, S.L., Kalvius, G.M., Snyder, R.E., Beard, G.B.: Phys. Rev. **148**, 176 (1966)
17. Ruby, S.L., Shenoy, G.K.: Phys. Rev. **186**, 326 (1969)
18. Ruby, S.L., Kalvius, G.M., Dunlap, B.D., Shenoy, G.K., Cohen, D., Brodsky, M.B., Lam, D.J.: Phys. Rev. **18**, 374 (1969)
19. Kalvius, G.M., Ruby, S.L., Dunlap, B.D., Shenoy, G.K., Cohen, D., Brodsky, M.B.: Phys. Lett. **29B**, 489 (1969)
20. Ruby, S.L.: J. Phys., Colloq. **C6**, 209 (1974)
21. Arthur, J., Brown, G.S., Brown, D.E., Ruby, S.L.: Phys. Rev. Lett. **63**, 1629 (1989)
22. Arthur, J., Brown, D.E., Ruby, S.L., Brown, G.S., Shenoy, G.K.: J. Appl. Phys. **67**, (1990)
23. Gerdau, E, de Waard, H. (eds.) : Hyperfine Interact. 123/124 and 125, Baltzer Science Publ., The Netherlands (1999)
24. Kishimoto, S.: Nucl. Instrum. Methods A **309**, 603 (1991)
25. Kishimoto, S.: Rev. Sci. Instrum. **63**, 824 (1992)
26. Baron, A.Q.R., Ruby, S.L.: Stanford Linear Accelerator Center Report, SLAC-PUB-6271, SLAC/SSRL-0038, October 1993
27. Baron, A.Q.R., Ruby, S.L.: Nucl. Instrum. Methods Phys. Rev. A **343**, 517 (1993)
28. Sturhahn, W.: Hyperfine Interact. **125**, 149 (1999)
29. Chumakov, A.I., Smirnov, G.V., Baron, A.Q.R., Arthur, J., Brown, D.E., Ruby, S.L., Brown, G.S., Salashchenko, N.N.: Phys. Rev. Lett. **71**, 2489 (1993)
30. Chumakov, A.I., Baron, A.Q.R., Arthur, J., Ruby, S.L., Brown, G.S., Smirnov, G.V., van Bürck, U., Wortmann, G.: Phys. Rev. Lett. **75**, 549 (1995)
31. Smirnov, G.V., van Bürck, U., Arthur, J., Popov, S.L., Baron, A.Q.R., Chumakov, A.I., Ruby, S.L., Potzel, W., Brown, G.S.: Phys. Rev. Lett. **77**, 183 (1996)
32. Perlow, G.J.; Phys. Rev. Lett. **40**, 896 (1978)
33. Seto, M., Yoda, Y., Kikuta, S., Zhang, X.W., Ando, M.: Phys. Rev. Lett. **74**, 3828 (1995)
34. Sturhahn, W., Tollner, T.S., Alp, E.E., Zhang, X., Ando, M., Yoda, Y., Kikuta, S., Seto, M., Kimball, C.W., Dabrowski, B.: Phys. Rev. Lett. **74**, 3832 (1995)
35. Alp, E.E., Sturhahn, W., Toellner, T.S., Zhao, J., Hu, M., Brown, D.E.: Hyperfine Interact. **144/145**, 3 (2002)
36. Sturhahn, W.: J. Phys., Condens. Matter **14**, S497 (2004)

Hyperfine Interact (2006) 170:15–32
DOI 10.1007/s10751-006-9475-2

Mössbauer spectroscopy methodology at the cutting-edge of high-pressure research

Moshe P. Pasternak · R. Dean Taylor

Published online: 10 January 2007

Abstract This paper provides a concise, introductory review intended mainly for Mössbauer spectroscopy (MS) scientists not familiar with the most modern aspects of High Pressure (HP) methodology. Following a short introduction to the 1st generation HP-MS based on Drickamer's pressure cells, we describe the principles of the 2nd generation of HP-MS based on the Diamond Anvil Cell (DAC) including *in-situ* pressure measurements, the use of the high-specific activity ^{57}Co(Rh) point sources, and examples of miniature DAC's. Finally, we present recent studies carried out with ^{57}Fe HP-MS combined with other HP techniques such as resistivity, and synchrotron-based x-ray diffraction, describing unique cases of the breakdown of magnetism and the Mott transition in hematite (Fe_2O_3), pressure-induced spin crossover in Wüstite (FeO), pressure induced $Fe^{2+} \rightarrow Fe^{3+}$ in $Fe(OH)_2$, and (P,T) induced inverse↔ normal spinel transition in magnetite (Fe_3O_4).

Key words ^{57}Fe Mössbauer effect · high-pressure · diamond-anvil cells · iron-oxides · Mott-transition · spin-crossover

1 Introduction

Pressure is one of the most effective methods available to solid-state scientists to alter properties of matter. Since its implementation by Bridgman early this century, application of high pressure research in matter has provided substantial information on the properties of matter, in all aspects,. Today, as a result of the development of static high pressure devices based on the *diamond-anvil cell* (DAC), experimenters can reach pressures in the megabar region generating *energy densities* in matter of the order of keV/nm^3. With such energy densities insulators with gap energies in the eV regions become metals, new structural and

M. P. Pasternak (✉)
School of Physics and Astronomy, Tel Aviv University, Tel Aviv, Israel
e-mail: moshepa@tau.ac.il

R. D. Taylor
MST-10, Los Alamos National Laboratory, Los Alamos, NM 87545, USA

electronic phases become stable, and new aspects of magnetism may be revealed. The DAC thus has become a very powerful ultra-high-pressure device, helping scientists discover new states of matter. Some of the modern DAC's generating pressures into the Mbar region can fit into the palm of the hand and allow a variety of sophisticated measurements to be performed on materials even though samples are of *microscopic* dimensions. The principles underlying the DAC, its pressure calibration, its applications as of 1983, and some potential uses are extensively described in the review paper by Jayaraman [1].

The goal of this paper is to introduce the Mössbauer spectroscopy (MS) investigator to the fascinating field of high pressure studies using the diamond anvil cell. As will be shown, the contributions of MS to research in the state of matter at very-high-pressures are significant. It is unique in its measurement of solids properties on the atomic scale and nicely complements X-ray, Raman, and IR spectroscopy in diamond anvil cells. MS is the only viable probe today for magnetic studies at pressures beyond 100 kbar. Due to the relatively large absorption cross sections, most Mössbauer isotopes have adequate sensitivity to probe the sub-milligram sized samples required in DAC's. In this article we assume the reader is familiar with the fundamentals of MS and with hyperfine interaction terminology, but not necessarily with DAC methods. After an introduction in Section 1, Section 2 is devoted to the methodology of the DAC emphasizing features relevant to MS, and Section 3 presents a few examples of ^{57}Fe HP-MS.

1.1 The first generation

About 3 years after the discovery of the Mössbauer effect [2, 3], Hanks published a short paper [4] – suggesting an application of Mössbauer spectroscopy (MS) to high pressure state of matter. The paper was entitled: *Pressure Dependence of the Mössbauer Effect.* Hanks apparently used a wrong formula to calculate the recoil-free fraction f for Cs but stated, "we find that a pressure of 5,000 atm will bring f to an (observable) value of 4×10^{-3}, i.e., will cause sufficient change in the crystal binding so as to change f by two orders of magnitude." Six months later Pound, Benedek, and Drever published [5] results of the first high pressure experiment with MS. Citing these authors: "The source, about 2 mC of Co^{57} diffused into iron, was enclosed in a beryllium–copper pressure bomb equipped with 1.5-in thick Be window. The bomb was cemented to the ferroelectric transducer and connected to a Bridgman press by a stainless steel tube." The pressure bomb used for this experiment is shown in Fig. 1. The highest pressure achieved was 3,000 kg/cm^2 (3 Kbar). The pioneering work in MS high pressure experiments of *the first generation* and the vast majority of published works, starting in early 1960s and into the mid-1970s, may be attributed to Drickamer et al. who also published several excellent reviews [6–10]. Although their studies were primarily conducted with the ^{57}Fe Mössbauer isotope either as sources or absorbers, their success, experience, and consequent progress in the high pressure methodology encouraged many other MS groups to initiate high pressure studies with a variety of isotopes. Effects of pressure on the electronic configuration of matter were studied with ^{57}Fe, ^{119}Sn, ^{125}Te, ^{153}Eu, ^{161}Dy, ^{170}Yb, ^{181}Ta, and ^{197}Au. An excellent review on *Mössbauer Studies at High Pressure* to 1975 has been published by Holzapfel [11].

In the *first generation* high pressure MS experiments the cells used were limited to pressures of ~20 GPa; however, the maximum pressure commonly was ~10 GPa (Abd-Elmeguid and Wortmann, private communications). This pressure is adequate for observing phase transitions in many of the covalent and some of the metallic materials (Abd-Elmeguid and Wortmann, private communications). Due to the bulky construction of the cells,

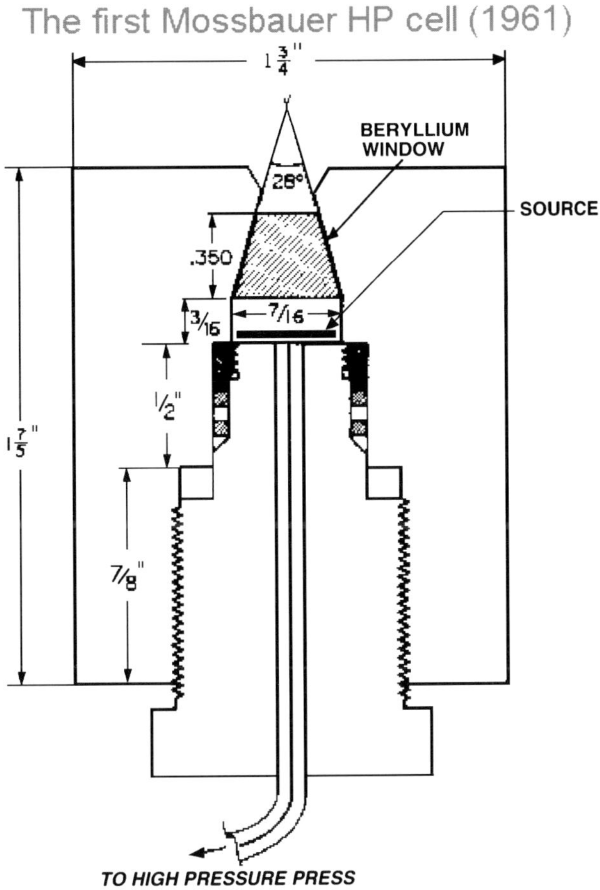

Fig. 1 A cross sectional view of the pressure cell (pressure bomb) used for the first MS experiment under pressure [5]

experiments at cryogenic temperatures and in external magnetic fields were not an easy task. For observation of structural and electronic phase transitions in oxides, minerals, and molecular crystals, pressures up to and beyond 100 GPa are required. Such ultimate hydrostatic pressures can be achieved only with diamond-anvil cells.

1.2 $P=F/A$

Pressure is obtained by applying a force on an certain area. One Newton applied on a square meter creates a pressure of one Pascal; 1 Pa=N/m^2 is the SI pressure unit. A common unit for expressing high pressure is the kilobar. 1 Kbar=10^3 bar=10^9 dynes/cm^2. In high pressure studies the gigapascal (1 GPa=10^9 Pa) is a more appropriate unit; 1 GPa= 10 kbar and 1 Mbar (10^6 bar)=100 GPa. The pressure at the center of the earth is estimated at 300 GPa. Using the ocean depths as a generator of *hydrostatic* pressure, one may achieve pressures of at most 0.1 GPa at a depth of 10,000 m! At this pressure the force exerted on 1 cm^2 corresponds to 1 metric ton.

It is obvious that the practical generation of *very* high pressures cannot rely on an ever larger external force, but rather on a reduced size of sample. Thus, to create hydrostatic pressures in the 10–100's GPa range at the laboratory one must induce uni-axial forces on areas as small as practical. One ton force exerted upon a 0.3×0.3 mm^2 area will generate pressures approximating 1 Mbar! To convey such forces to the samples one needs extremely hard materials that will preserve their elastic properties even at this pressure. High quality crystalline diamond anvils, the hardest substance known to man, are quite transparent to X-rays and light and are the only present solution for such anvils. Anvils made of 0.20–0.35-carat diamonds are adequate, and their prices are affordable. Finally, to transform the uni-axial pressure into a hydrostatic pressure one needs to contain the samples within hard metal gaskets. The main factors that contributed to the significant advances in high pressure research in recent years are attributed to the development of diamond anvils, proper gasketing materials and hydrostatic transmitting media, progress in powerful miniature sources of electromagnetic radiation at many wave-lengths, and the availability of sensitive spectrometers. MS is the most recent laboratory-operated spectroscopy to be employed using DACs, herein referred to as MS-DAC.

2 The second generation

2.1 Principles of the diamond anvil cell

The basic configuration of all diamond anvil cells is shown in Fig. 2. The diamonds (1) modified to have flat culets (3) are pressed into the sample, which is confined in the sample chamber (5) formed in a preindented metal gasket (4). A closure force is applied to the tables (2) via appropriate backing plate and closure assemblies. Numerous styles of DACs have been developed, and many of them carry the names of the laboratories, e.g., the *NBA cell* [12] or of the researchers, e.g., the *Mao–Bell cell*, [13] the *Holzapfel cell* [1], or the *Merrill–Bassett cell*. [14] DAC's can be classified into two main types: (1) – *Opposed-plate* cells and (2) – *Piston-cylinder* cells. Whereas cells of the first class are relatively simple, inexpensive, easy to handle, and suitable for pressures up to ~60 GPa (with say 300-μm culets), cells of the second class are more robust, more difficult to align, more precise in its closure, and more expensive, but they are suitable for pressures up to and beyond 100 GPa when using culets sizes of about 150 μm or smaller. Cells typical of the opposed-plates class are shown in Fig. 3 and one of the piston cylinder classes [15] is shown in Fig. 4. The triangular Merrill–Bassett cell, designed for XRD but also used for MS, is the simplest working extension of the basic configuration and is shown in cross-section in Fig. 3 along with a schematic representation of the MS source (5), absorber (8), and detector. The sample (8), ruby chips (7), and hydrostatic medium are placed in a hole drilled in the preindented gasket material (6) placed between the flat parallel faces (culets) of two opposed anvils (4) and are subjected to pressure when squeezed by the three screws acting on the backing plates (2) via the cell assembly plates (1). Details are given below.

Diamonds Diamonds for anvils are usually selected from flawless, gem-quality stock. The *brilliant* cut diamonds are especially faceted with 16 sides to provide a more nearly circular culet face. Anvils are available from commercial sources [16]. Sizes vary from 0.2–0.4 carat depending on the particular need. The culets are cut and polished parallel to the table (see Fig. 2) to a very high degree of accuracy. For ^{57}Fe MS it is advisable to apply partially perforated anvils [17].

Fig. 2 A generic view of a diamond anvils cell

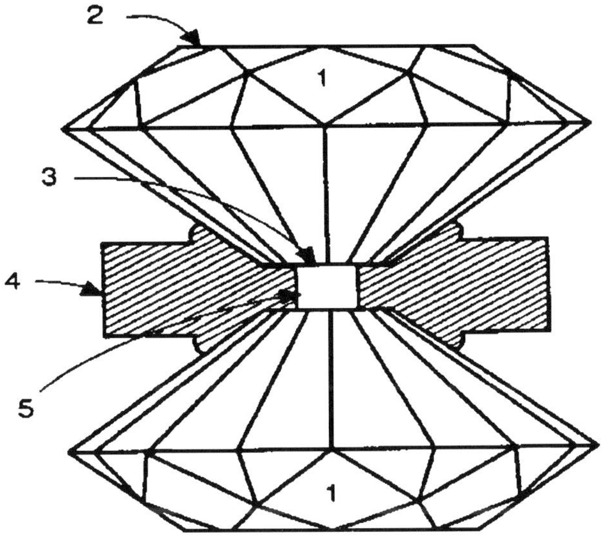

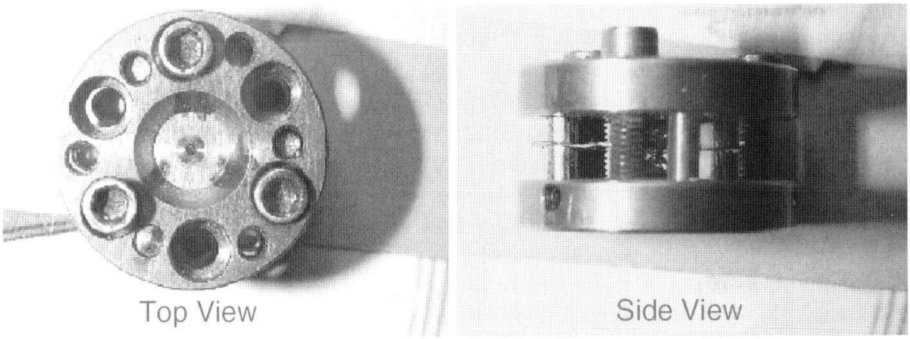

Fig. 3 An example of an Opposing-plates DAC [69]

Gasketing The use of a gasket for the containment of pressure was first demonstrated by Van Valkenburg. [18] This significant discovery allowed using the DAC as a quantitative tool for high pressure research. Besides providing for hydrostatic sample containment, the gasket acts as a supporting ring by extruding around the diamond culets, reducing failures of the anvils due to lateral stress and edge stress. Gaskets are for MS also serve as collimators by using high-Z materials such as $Ta_{90}W_{10}$ or Re.

Manometry Pressure calibration in DAC's is performed by the ruby fluorescence method [19]. This method based on the quasi-linear pressure dependence of ruby (Cr^{3+}-doped Al_2O_3) *R*-lines – 692.7 and 694.2 nm – simplifies enormously the in-situ determination of pressure in a DAC and its distribution. The ruby *R* lines are excited by blue or green light (Cd–He or Ar–ion) lasers, and the fluorescence wave-lengths are measured with a standard monochromator. The pressure dependence of the wavelength shift $\Delta\lambda$ has been

Fig. 4 An example of a Piston-cylinder DAC [59]

experimentally calibrated using the established equation of states of several elements [20]. The expression suitable for pressures to at least 100 GPa is given as:

$$P = 0.03808\left[(\Delta\lambda/694.2 + 1)^5 - 1\right], \tag{1}$$

where P is in GPa and $\Delta\lambda$ in nm.

2.2 A brief history

The first reported high-pressure (15 GPa) MS done with diamond anvils was in 1965 by Herber and Spijkerman [21]. Measurements were carried out in $^{119}SnO_2$ using a conventional NBA cell modified for diamond anvils. Presumably no gasket was used, and no description is provided about the pressure calibration or sample confinement. A short report by Huggins et al. [22] on a MS-DAC study appeared in *Carnegie Institution Washington Year Book,* 74. However, the crucial modern studies that led into the present state of art of MS-DAC were effectively initiated in the early 1980s at Los Alamos with Fe (^{57}Co) source experiments [23] to P=21 GPa and at the University of North Carolina with β-Sn absorber studies [24, 25] to P=30 GPa. The first International Conference on the Applications of the Mössbauer Effect at which MS-DAC studies were presented was ICAME-1985, Leuven. Papers by Pasternak et al. [26] dealt with studies of $^{129}I_2$ molecular crystals to P=35 GPa at cryogenic temperatures and by Nasu et al. [27] with ε-Fe and Fe_2O_3 at room temperature to P=45 GPa and 76 GPa, respectively. These early publications with ^{57}Fe, ^{129}I and with ^{151}Eu [28] motivated many of the existing high-pressure MS laboratories either to switch or to add a DAC capability to their facilities.

3 Examples of recent ^{57}Fe HP–MS

3.1 Breakdown of strong correlation; the case of Hematite (Fe_2O_3) [29]

Fe_2O_3, a wide-gap antiferromagnetic insulator (T_N=956 K [30]) with its large optical gap can be regarded as archetypal of a *Mott*-insulator. Motivated primarily by its importance in the earth sciences, high-pressure studies of this corundum-type mineral have been extensively performed since the early days of high-pressure physics. As early as in the

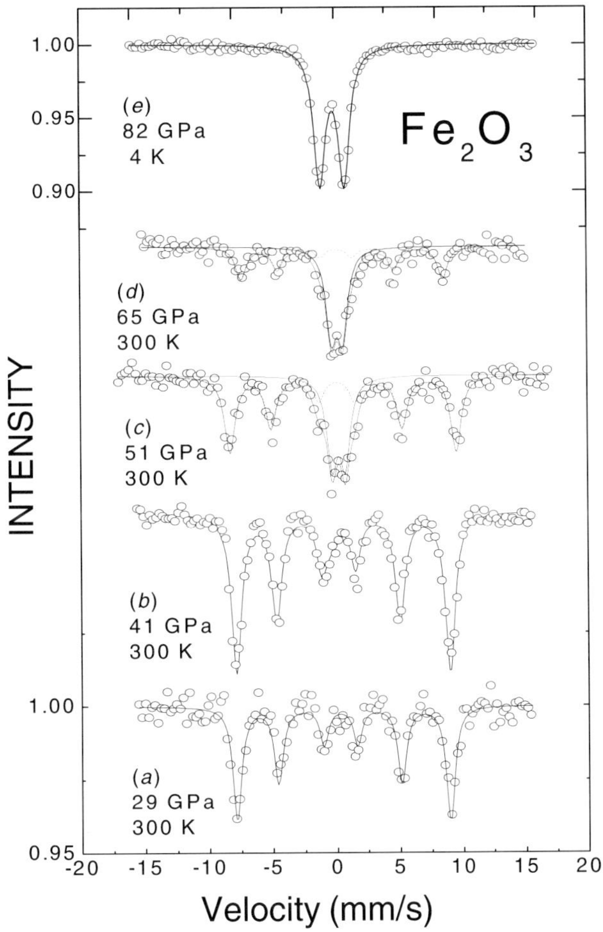

Fig. 5 Mössbauer spectrum of Fe_2O_3 at various pressures at 300 K. The spectrum at 82 GPa was measured both at 300 K and 4 K

mid-1960s, shock-wave experiments were performed by McQueen and Marsh [31] who reached energy densities not available at that time in conventional static pressure cells. Based on those experiments Reid and Ringwood [32] proposed a new denser structure formed in the 60–120 GPa range. In the mid-1980s, static high pressure conventional XRD studies carried out with DAC's by Yagi and Akimoto [33] and Suzuki et al. [34] confirmed the onset of a new denser phase at ~50 GPa. This structure could be ambivalently assigned either to the "distorted corundum," Rh_2O_3-II-type [35] or to an orthorhombic perovskite structure. Toward the late 1980s with the advent of synchrotron radiation facilities and their availability for materials science studies, XRD measurements using the energy-dispersive mode were carried out to 67 GPa by Staun Olsen et al. [36]. They reported the onset of a first-order phase transition at ~50 GPa accompanied by a drastic volume reduction of about 10% and assigned an orthorhombic structure to this new phase. Yet, despite improved experimental data of the XRD patterns, the authors did not unambiguously differentiate between the Rh_2O_3-II corundum-type or the *perovskite* structures.

In this study we have performed extensive measurements, up to 80 GPa, with XRD in the angle-dispersive mode [37], Mössbauer Spectroscopy (MS), and Resistivity employing the TAU miniature piston/cylinder DAC [38] with anvils having 0.3-mm flats. Ruby

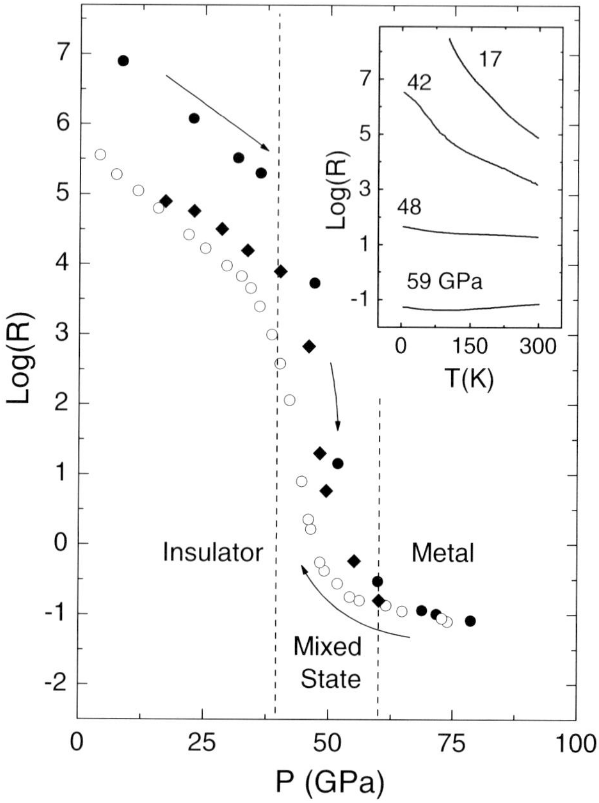

Fig. 6 The *R*(*P*,*T*) *curve* of hematite showing clearly the insulator-metal transition driven by correlation breakdown at high-pressures

fluorescence was used for pressure measurements. For XRD and MS [39] samples of spectroscopic pure Fe_2O_3 were encapsulated in 150-μm diameter by 25-μm high cavities drilled in SS and Re gaskets, respectively. Argon was used as a pressurizing medium. For resistivity studies the ring surrounding the sample cavity drilled in a SS gasket was coated with fine Al_2O_3 powder mixed with epoxy for the purpose of electrical insulation of the 5-μm thick Pt electrodes, and the four-probe method was used. The average distance between electrodes was 40 μm, and the pressure distribution pertinent to the resistance measurements was typically within 5% of the average pressure. At selected pressures, particularly at the vicinity and above the phase transition, the temperature dependencies of *R*(*P*) and of MS were carried out in the range 4–300 K using a *dip-stick* sample holder immersed into a He-storage dewar and a top-loaded cryostat [40], respectively.

Mössbauer spectra of Fe_2O_3 characteristic of various pressure ranges taken at 300 K are shown in Fig. 5 a–d. Up to ~45 GPa the dominant spectral component is that of the LP phase (a, b), arising from the $^6A_{1g}$ high-spin state and characterized by H_{hf}=51 T; a typical value of the hyperfine field for ionic ferric oxide bonding. In the 50–55 GPa range a non-magnetic quadrupole-split component emerges (c), designated as the HP component, *coexisting* with the 51 T magnetic-split LP-component. In the 60–70 GPa range (d) the relative abundance of the LP component keeps on decreasing, and we note that the H_{hf} at 300 K of the magnetic component is slightly reduced, by about 10%; this magnetic phasein the mixed region is designated as an intermediate phase (IP). For *P*>72 GPa, the only spectral component observed in the 4–300 K range is that of the non-magnetic HP phase (e)

Fig. 7 The XRD pattern of Fe_2O_3 showing the transition from a normal corundum phase (space group) at low pressures to a Rh_2O_3-II-type, with space group Pbna

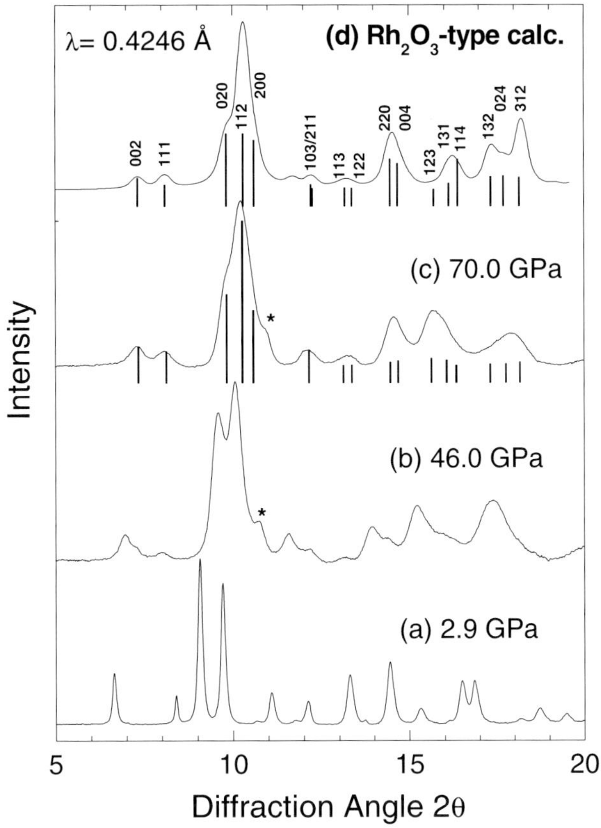

and is characterized by a quadruple-split spectrum. The lack of a magnetic ordering down to 4 K prompted us to conclude that this single HP component reflects not a paramagnetic, but rather a non-magnetic state [41]. This signals the breakdown of the *d–d* electronic correlation, and that will be accompanied by metallization.

The pressure variation of the resistance, *R*(*P*), at 300 K is shown in Fig. 6. As can be seen a precipitous decrease occurs at the onset of the phase transition in which *R*(*P*) is reduced by more than six orders of magnitude [42]. The resistivity value at 80 GPa and 300 K was estimated to be $1.5(7)\times10^{-6}$ Ωm, typical of a metal. To further explore the HP electronic state we carried out *R*(*T*) measurements at various pressures (see inset in Fig. 2). As shown, the *R*(*T*, 17 GPa) curve for the LP regime is typical of an insulator, and the *R*(*T*, 59 GPa) curve with its positive d*R*/d*T*, is characteristic of a metallic state. In the range 42–59 GPa the sample behaves as a mixture of insulating and metallic states with the metallic contribution (HP abundance) increasing with pressure. This trend is in full agreement with the MS results (see Fig. 1) provided we assign non-magnetic and magnetic components to the metallic and insulating phases, respectively.

The diffraction patterns typical of the LP, IP, and HP regimes are shown in Fig. 7. The LP diffraction pattern (a) fits very well with the corundum phase (space group $R\,\overline{3}\,c$). The diffraction lines corresponding to the HP phase first appear at ~45 GPa, and the diffraction pattern recorded at 46 GPa (b) is characteristic of IP the intermediate regime where both LP and HP phases coexist. Diffraction patterns (c) of the pure HP recorded at 70 GPa can be precisely indexed to the distorted corundum phase, Rh_2O_3-II-type, with space group Pbna (see Fig. 3d).

This series of experimental data, using XRD, MS, and resistance measurements determined unequivocally the non-magnetic metallic state of hematite induced by a precipitous volume decrease at 50 GPa. The mechanism responsible for the insulator-metal transition is the breakdown of the strong *d–d* correlation resulting in a non-magnetic metal. The insulating and metallic phases coexist in the 42–58 GPa pressure range. Also, the ambivalence regarding electronic state of Fe and the HP structure is now clarified. The structural transition is minor, from the corundum-type to a distorted corundum Rh_2O_3-II type, with both composed of Fe^{3+} cations six-fold coordinated to oxygens. The geological implications of this finding are quite compelling. At depths corresponding to the inner mantle, hematite is metallic with a resistivity as high as that of metallic iron.

3.2 The high-spin to low spin transition; the case of Wüstite (FeO) [43]

Details of the state of matter at very high static densities are of fundamental significance to quantum mechanics, to the nature of magnetic ordering, and to band structure of narrow d-band materials. At the substantial high pressures that are now available with diamond anvils cells (DAC) and appropriate spectroscopical methods, new insights on magnetic properties are emerging. Magnetic properties are studied preferably in simple binary transition-metal (TM) compounds, the so-called *Mott Insulators* [44] which, by virtue of the strong on-site electron-electron correlation within the d-bands have localized moments, become antiferromagnetically ordered at temperatures below T_N and are insulators with large optical gaps that persist at temperatures far above T_N. In the isostructural regime of density increase, the magnetic state will, in principle, eventually collapse as a result of one of the following mechanisms:

(1) Insulator-metal transition resulting from the closure of the Mott–Hubbard *d–d* gap (U) or of the Charge–Transfer *p–d* gap (Δ). Experimental verification of such a phenomenon is the concurrent metallization *and* collapse of magnetic moments within a narrow pressure range. The pressure-induced so-called *Mott transition* has been experimentally demonstrated in the case of NiI_2 by Pasternak et al. [45] and later on in CoI_2 [46] and FeI_2 [47] using the combined methods of Mössbauer Spectroscopy (MS), X-ray diffraction (XRD), and electrical conductivity in DAC's.

(2) High-spin (HS) to low-spin (LS) transition resulting from the breakdown of Hund's rule at very high density. In HS TM-compounds a transition to LS will occur when the *crystal field* splitting exceeds the *exchange energy*, and the material will become diamagnetic resulting in the collapse of magnetic state, for the particular case of an even valence number n in a d^n configuration.

This letter an experimental proof has been reported for such an isostructural/isochoric HS→LS transition [48] in wüstite ($Fe_{1-x}O$). This has been observed with ^{57}Fe MS used as a probe of magnetism at pressures to 120 GPa. To our knowledge this is the first report on Mössbauer Spectroscopy beyond 100 GPa.

Wüstite is non-stoichiometric divalent (d^6) iron oxide invariably with a cation deficiency (x). Depending on x, T_N at ambient pressure may vary between 190–210 K in the $0.90<\times<0.96$ range [49]. It crystallizes in a NaCl structure at ambient pressure and at ambient temperature distorts into a rhombohedral cell above ~18 GPa. In both structures Fe

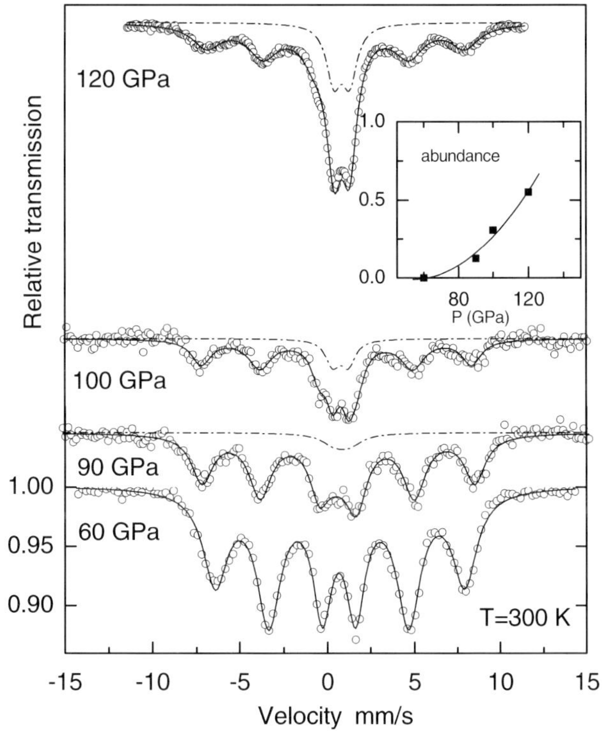

Fig. 8 Typical Mössbauer spectra of $Fe_{0.94}O$ measured at $P \geq$ 60 GPa at 300 K. The *solid line* is a least-square-fit to the experimental points assuming magnetic- and quadrupole-split components with varying intensities. The *inset* depicts the variation of the relative abundance of the non-magnetic component (the LS-component, see text) with pressure. Note that by extrapolation the HS-component will be fully converted to the LS at ~140 GPa

is at the center of an oxygen octahedron. A modified piston/cylinder Mao–Bell type DAC was used with anvils having culet flats of 150 μm. A 75-μm hole was drilled in a Re gasket into which the sample was loaded in a 25 μm high cavity along with ruby chips for pressure determination. XRD using synchrotron radiation was also used for independent determination of pressure. Most measurements were carried out using a top-loading cryostat with experiments conducted in the 4–300 K range.

Typical spectra at 300 K obtained at 60, 90, 100 and 120 GPa are shown in Fig. 8. With increasing pressure, at ambient temperature, a new non-magnetic (nm) component evolves at ~90 GPa, and its abundance increases with pressure as shown in the inset of Fig. 1. At 120 GPa this spectral component is characterized by a quadrupole splitting (QS) and an isomer shift (IS) of 0.85(7) mm/s and 0.75(10) mm/s, respectively. Those values are higher than those obtained for the magnetic component (QS=0.2(2) mm/s and IS=0.5(1) mm/s). The *nm* abundance is also temperature dependent. A typical case is shown in the inset of Fig. 3 where the presence of the *nm* component at 120 GPa is first detected at T>70 K, and its abundance increases with temperature. At this pressure full conversion is extrapolated to be ~450 K. These and all other measurements showed both temperature and pressure reversibility; no significant hysteresis was observed. We assign this *nm* spectral component to the $^{1}A_{1g}$ diamagnetic low-spin state of Fe^{2+}.

This has been the first manifestation of Mössbauer spectroscopy studies beyond 100 GPa. We found that the collapse of the magnetic state in $Fe_{1-x}O$ is due to a constant-volume (probably second-order) HS→LS transition, where both species coexist over the pressure range of 90–140 GPa at $T \leq 300$ K. The energy gap separating the diamagnetic $^{1}A_{1g}$

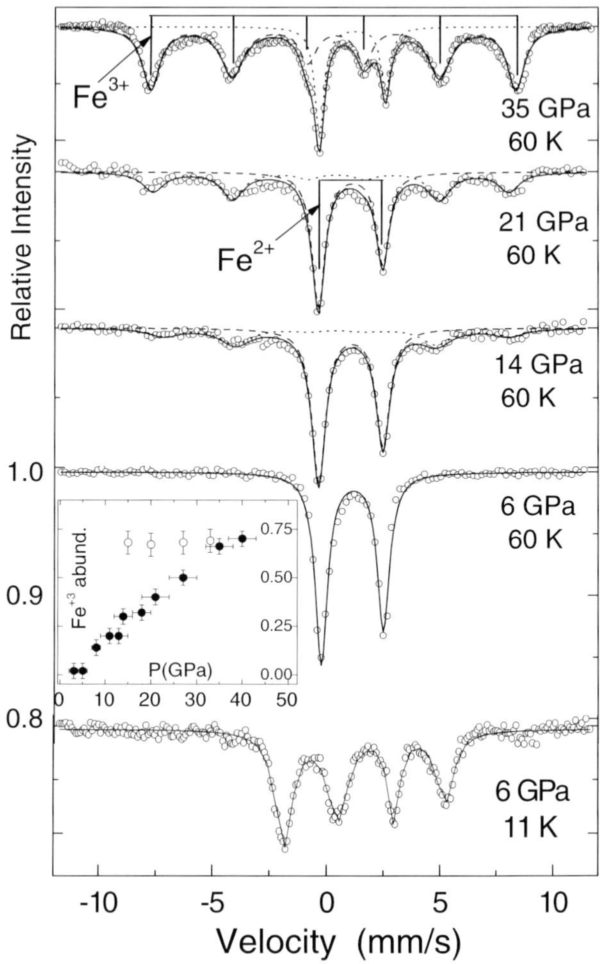

Fig. 9 Mössbauer spectra of $^{57}Fe(OH)_2$ for different pressures measured at 70 K. The spectrum measured at 6 GPa, 70 K corresponds to the paramagnetic state characterized by a QS=2.7 and IS=1.0 mm, typical ferrous compounds. The slight asymmetry in the doublet intensities is caused by texture effects. The asymmetry increases with pressure. The spectrum measured at 6 GPa and 11 K shows the magnetically ordered state hyperfine splitting. The *solid line* is the theoretical spectrum calculated from the full spin-Hamiltonian governing the hyperfine interaction (see text) resulting in H_{hf}=19 T, QS=2.6 mm/s and IS=1.0 mm/s. At P >6 GPa the Mössbauer spectra reveal the onset of a new component corresponding to a ferric species whose sublattice ordering temperature is considerably higher than that of the ferrous species. With increasing pressure (14–35 GPa) the relative abundance of the Fe^{3+} component increases reaching ~70% at 40 GPa (see inset). The hyperfine parameters of the ferric component obtained were: H_{hf}(70 K)=50 T, QS(70 K)=0.25 mm/s and IS(300 K)=0.25 mm/s

and magnetic $^5T_{2g}$ states decreased with pressure, reaching values comparable to ambient temperature k_BT at ~100 GPa.

3.3 Pressure-induced "oxidation" of Fe^{2+}; the case of iron hydroxide ($Fe(OH)_2$) [50]

Structural properties of $M(OH)_2$ under static pressure have been the subject of numerous studies during the last few years. The main motivation was an early experimental discovery by Kruger et al. [51] of a pressure-induced broadening of the O–H vibrational A_{2u} modes in both $Mg(OH)_2$ and $Ca(OH)_2$, observed by way of infrared spectroscopy. The broadening phenomenon was later confirmed by Duffy et al. [52] from high-pressure Raman studies in $Mg(OH)_2$. A similar phenomena was later observed by Nguyen et al. [53] in $Co(OH)_2$ using both Raman and IR spectroscopy. From X-ray diffraction studies in the $Co(OH)_2$ case, the authors attributed the broadening to a local "hydrogen sublattice amorphization," with no effect on the long-range ordering of the Co–O lattice. The several studies have shown that the O–H vibrational broadening first appears at ~11–14 GPa, increases monotonically with pressure, and is reversible upon decompression.

The CdI_2 related hydroxides (space group $P\bar{3}m1$) have a rather unique structure with hydrogen atoms isolated between layers of metal ions in octahedral coordination with oxygen. By applying pressure one may follow a series of controlled O-H–H and O-H–O stages of repulsion and attraction, respectively. And indeed, a recent systematic high pressure (HP) study by Parise et al. [54] in $Co(OD)_2$ using neutron powder diffraction has shown that the hydrogen atoms become continuously disordered with increasing pressure as a result of structural frustration due to interlayer H–H repulsion. In the CdI_2-related structure, at ambient pressure, M is located at (0,0,0) and O and H at (1/3, 2/3, z) with ~0.2 for O and ~0.4 for H. At ambient pressure the O-H axis is aligned along the three-fold symmetry c-axis of the crystal, but because of the pressure-induced repulsion, the $(O\text{-}H)^{-1}$ dipole axis "splays" with respect to the c-axis direction with the bending angle increasing with pressure (see Fig. 9). According to [54], the H position is continuously laterally displaced from its original three-fold axis, reaching D=0.4 Å at 16 GPa. The pressure-induced H-displacement or $(O\text{-}H)^{-1}$ dipole *bending* is responsible for the vibrational broadening as observed by optical spectroscopy in all $M(OH)_2$ HP studies. As will be shown this dipole orientation mechanism is also responsible for the continuous self-oxidation, or ionization, of the ferrous ion in $Fe(OH)_2$ [55, 56].

Typical Mössbauer spectra measured at 6 GPa, 11 K ($T<T_N$) and at 60 K ($>T_N$) at various pressures are shown in Fig. 9. Below ~8 GPa the spectra are composed of a single Fe^{2+} component characterized by quadrupole splitting QS($e^2q_{zz}Q$)=2.95(2) mm/s and isomer shift IS=1.00(1) mm/s, with values barely changing to 40 GPa. The asymmetry in intensity of the QS doublet which increases with P is attributed to a sample texture effect. The hyperfine field H_{hf} obtained from the least-squares-fitting at 11 K is H_{hf}=19.5 (2) T oriented at 90° (270°) relative to the principal axis of the electric-field-gradient, eq_{zz}. In the 0–40 GPa range T_N increases monotonically, reaching 63 K at 40 GPa. Spectra taken at $P>$ 8 GPa show the presence of a new, additional component whose relative abundance increases with pressure. The hyperfine parameters of this new component are: IS=0.3 (1) mm, QS=0.3(1) mm/s and H_{hf}=51(1) T values characteristic of Fe^{3+} in an oxygen six-fold coordination configuration. The relative abundance of the Fe^{3+} component was evaluated using the appropriate areas A of the respective absorption spectra components:

$$\mathrm{Fe}^{3+}\,\text{abundance} = \frac{A(\mathrm{Fe}^{3+})}{A(\mathrm{Fe}^{3+}) + A(\mathrm{Fe}^{2+})} = \frac{f_{\mathrm{Fe}^{3+}}\, n_{\mathrm{Fe}^{3+}}}{f_{\mathrm{Fe}^{3+}}\, n_{\mathrm{Fe}^{3+}} + f_{\mathrm{Fe}^{2+}}\, n_{\mathrm{Fe}^{2+}}}$$

The continuous increase of H disorder resulting in the formation of splayed dipoles and possibly multipoles of O-H units would no doubt produce a finite electric field at the Fe site. And such an induced potential must be sufficient to remove this weakly bonded electron creating either a new ferric hydroxide $Fe^{3+}O(OH)$:

$$\mathrm{Fe}^{2+}(\mathrm{OH})_2 \xrightarrow{\mathrm{P}} \mathrm{Fe}^{3+}\mathrm{OOH} + \frac{1}{2}H_2 \tag{2}$$

or a new electronic-band within the high-pressure band structure:

$$\mathrm{Fe}^{2+}(\mathrm{OH})_2 \xrightarrow{\mathrm{P}} \mathrm{Fe}^{3+}(\mathrm{OH})_2 + e^- \tag{3}$$

To show whether a new ferric compound is created (expression (2)), coexisting with the ferrous hydroxide (and a free H_2), we carried out a series of HP-XRD to 30 GPa. Careful XRD analyses have shown that the indexes of $Fe(OH)_2$ persist to 38 GPa.

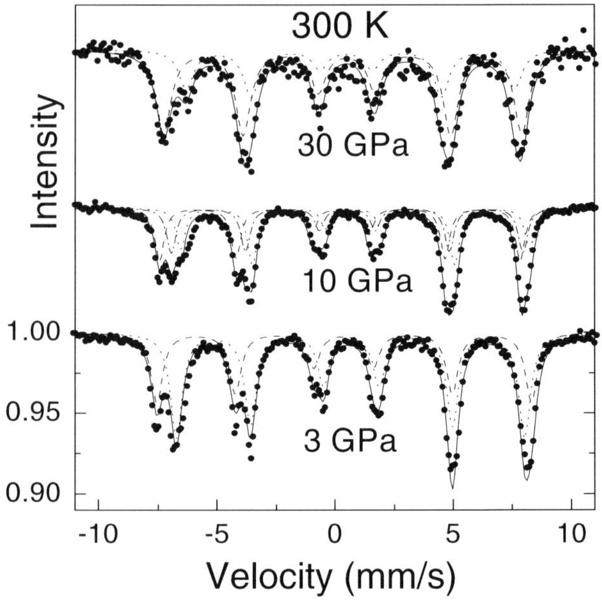

Fig. 10 Mössbauer spectra recorded at 300 K for various pressures. With increasing pressure distinct changes in the spectral components are observed. The 3 GPa and the 30 GPa spectra correspond to the inverse and normal spinel, respectively. The 10 GPa spectrum corresponds to the decoupled d-charge at the inverse-spinel state. The *dashed curve* (-) corresponds to Fe^{3+} at the A-site, the *dashed-dotted* (-·-·-·-·-) curve to Fe^{3+} in B-site and the dotted (········) curve to Fe^{2+} in the B-site

These series of experimental data, using XRD, MS, and R measurements at various pressures and temperatures determined unequivocally the gradual pressure-induced oxidation of Fe^{2+} to Fe^{3+}. This phenomenon of "self-oxidation" in $Fe(OH)_2$ is attributed to the gradual lateral displacement of the H atom, due to coulomb repulsion, that culminates in the formation of *splayed* (*O-H*) dipoles. The breakdown of the (*O-H*) axial symmetry creates an effective dipole potential at the iron site in addition to the relatively weak binding energy of the last valence-electron contribute to the ionization of the Fe^{2+}. This self-oxidation does not lead to a new ferric-hydroxide species, e.g., $Fe^{3+}O(OH)$, but rather to the creation of a new electron-band within the HP band-structure of $Fe(OH)_2$. Up to ~50 GPa those electrons are rather localized, but at $P>50$ GPa a first-order electronic transition takes place in which the electrons become weakly delocalized and the resistivity decreases discontinuously by ~2.5 orders of magnitude.

3.4 The coordination crossover in magnetite (Fe_3O_4); a pressure-induced inverse-to-normal spinel transition [57]

Following the discovery by Verwey [58], of a remarkable discontinuous drop in the conductance upon cooling below $T_V=122$ K, the *Verwey temperature*, magnetite became the subject of numerous studies in condensed matter physics. This peculiar electronic transition occurs at or close to a first-order structural phase transition from a cubic to a monoclinic structure [59]. At $T>T_V$ magnetite has an *inverse* spinel structure (cubic crystal symmetry), in which the tetrahedral sites (*A*-sites) are occupied by Fe^{3+} and the octahedral sites (*B*-sites, twice as abundant as *A* sites) by Fe^{2+} and Fe^{3+}. The presence of the mixed-valence Fe ions in the *B* sites motivated Verwey et al. [60], including Mott [61], to hypothesize that electronic exchange between the ferrous and ferric ions take place above T_V. Charge delocalization would then explain the metallic behavior of magnetite at $T>T_V$

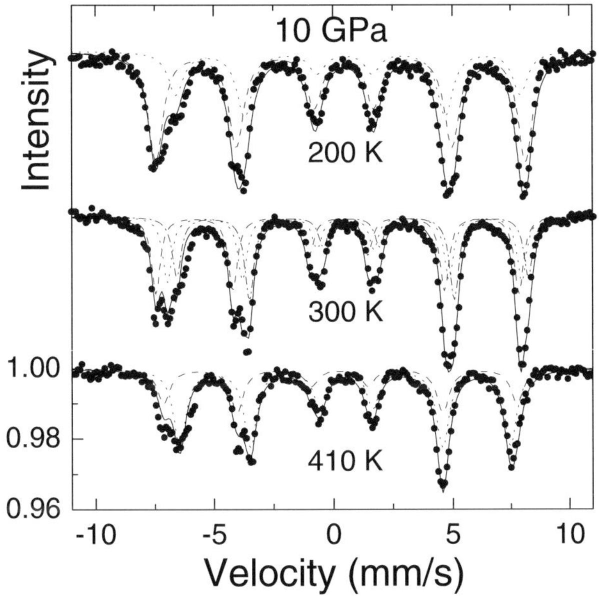

Fig. 11 Mössbauer spectra recorded at 10 GPa for 200, 300, and 410 K. The parameters were free to vary during the least-squares-fitting procedure. At 300 K, the abundance of the three components is equal within the experimental error. The abundance of the two components at 200 K (normal spinel) and at 410 K (inverse spinel) was 2/1, within the experimental error. Typical values of the deduced hyperfine fields at 200 K are 48.4 and 45.1 T for the $B(Fe^{3+})$ and A (Fe^{2+}) sites, respectively, and at 410 K they are 46.2 and 50.2 T for the $A(Fe^{3+})$ and $B(Fe^{2.5+})$, respectively

whereas the onset of charge localization at the respective individual iron ions below T_V would explain the metal-insulator transition.

Without question the need for an experimental proof of the suggested transition mechanism:

$$\left[Fe^{3+}\right]_A\left[Fe^{2+}Fe^{3+}\right]_B \xrightarrow{T\ >\ T_V} \left[Fe^{3+}\right]_A\left[Fe^{2.5}Fe^{2.5}\right]_B$$

is essential. Spectroscopic methods offer a useful means to probe directly the charge state of the Fe constituents. Though some spectroscopic methods such as neutron and resonant X-ray scattering are both atom- and site-specific probes, Mössbauer spectroscopy (MS) is indeed a preferred method because of its capability to differentiate between *iron charge-states* via the observed ^{57}Fe hyperfine parameters that include the isomer shift (IS), quadruple splitting (QS), magnetic hyperfine field (H_{hf}), and *relative abundance* of the spectral components.

None of the MS studies[6] carried out at $T<T_V$ could unequivocally resolve the elusive charge localization at the B-sites. A typical spectrum for localization would be composed of *three* distinct components, each with the same intensity: two components corresponding to the $Fe^{3+}Fe^{2+}$ pair on the octahedral site and the third to Fe^{3+} on the tetrahedral site. Yet, a typical Mössbauer spectrum below T_V is composed of only two components. Based on ^{57}Fe NMR, Novák et al. [62] concluded that no $Fe^{3+}Fe^{2+}$ ordering could be present below T_V. Similar conclusions were obtained by Garcia et al. [63] using the method of X-ray resonant scattering of the forbidden (0 0 2) and (0 0 6) reflections at the Fe *K* edge. As in the case of $T>T_V$ they concluded that both of the B-site Fe-charges must be identical.

Mössbauer spectra taken at RT and 3, 10, and 30 GPa, are shown in Fig. 10, and spectra corresponding to 10 GPa recorded at 200, 300, and 410 K are shown in Fig. 11. Features of these two sets of spectra are very similar. The spectrum at 3 GPa and RT (Fig. 10) is typical of those measured at ambient pressure and $T>T_V$ (see Fig. 1, 130 K). These spectra are characterized by two major components with an abundance ratio of 2/1, the one with higher abundance attributed to the mixed charge state $Fe^{2.5+}$ $Fe^{2.5+}$ in the *B* (octahedral) sites and the second component to Fe^{3+} at the *A* (tetrahedral) sites. At 30 GPa the spectrum looks like the 108 K spectrum in Fig. 1, characteristic of spectra at $T<T_V$ at ambient pressure.

In conclusion Mössbauer studies carried out by many authors at $T \ll T_V$ have produced rather complex spectra which apart from the two main components as seen at 108 K in Fig. 1, were fitted with up to five sub-spectra [64]. Many authors *assume* an inverse-spinel structure at $T<T_V$ and have therefore interpreted their results in terms of localized charges, with Fe^{2+} and Fe^{3+} valence states in the B-site. It should be pointed out that at low temperatures; strains have been documented in XRD studies [65], so the presence of several extra Mössbauer sub-spectra is not too surprising. A similar case is observed in FeO at low temperatures [66]. These lines are suppressed with pressure both in FeO and all of our high pressure spectra at low temperatures and could well be fitted with two components. It should be emphasized that many of the recent Fe_3O_4 studies and their analysis in particular, were *model-sensitive*. All analyses have assumed the existence of inverse-spinel structure bellow and above T_V. The present results and interpretation *are consistent* with both the NMR and the X-ray resonant scattering observations, which are model-insensitive, that reveal identical charge states for the two Fe in the B-site both above and below T_V, e.g., $Fe^{+2.5}$ and Fe^{+3}, respectively.

Thus, by virtue of the inverse to normal spinel transition the *d* electrons become localized, the Fe^{3+} and Fe^{2+} in the *B* and *A* sites, respectively, this cannot be accounted for the metal-insulator transition at T_V. A possible explanation for the Verwey transition at $T=T_V$ is the opening of a gap in the ligand *p*-band, as in the case of $Sr_3Fe_2O_7$ [67, 68]. The intra-band *p–p* gap does not involve the d-bands explicitly, as the latter is still separated by the Hubbard–Mott energy *U*. This explanation is consistent with the retention of the magnetic properties of magnetite both below and above T_V.

References

1. Jayaraman, A.: Rev. Mod. Phys. **55**, 65 (1983)
2. Mössbauer, R.L.: Z. Phys. **151**, 124 (1958)
3. Mössbauer, R.L.: Naturwissenshaften **45**, 538 (1958)
4. Hanks, R.V.: Phys. Rev. **124**, 1319 (1961)
5. Pound, R.V., Benedek, G.B., Drever, R.: Phys. Rev. Lett. **7**, 405 (1961)
6. Drickamer, H.G.: In: Seitz, F., Turnbull, D. (eds.) Solid State Physics, vol. 17, p. 1. Academic, New York (1965)
7. Drickamer, H.G.: Chem. Br. **9**, 353 (1973)
8. Drickamer, H.G., Frank, C.W.: Ann. Rev. Phys. Chem. **23**, 39 (1972)
9. Drickamer, H.G., Vaughan, R.W., Lewis, G.K. Jr.: Comments Solid State Phys. **1**, 163 (1968)
10. Frauenfelder, H., Ingalls, R.: In: Applications of Mössbauer Effect in Chemistry and Solid State Physics. Technical Report Series, no. 50, p. 37, International Atomic Energy Agency, Vienna (1966)
11. Holzapfel, W.B.: CRC Critical Reviews in Solid State Sciences, p. 89, (1975)
12. Piermarini, G.J., Block, S.: Rev. Sci. Instrum. **46**, 973 (1975)
13. Mao, H.K., Bell, P.M.: In: Carnegie Institution of Washington Year Book **77**, 824 (1979)
14. Merrill, L., Bassett, W.A.: Rev. Sci. Instrum. **45**, 290 (1974)
15. Machavariani, G.Yu., Pasternak, M.P., Hearne, G.R., Rozenberg, G.Kh.: Rev. Sci. Instrum. **69**, 1423 (1998)
16. D'Anvils Ltd. http://www.danvils.com (A company specially devoted to anvils, DACs and accessories for Mössbauer spectroscopy)
17. Dadashev, A., Pasternak, M.P., Rozenberg, G.Kh., Taylor, R.D.: Rev. Sci. Instrum. **72**, 2633 (2001)
18. Van Valkenburg, A.: Conference Internationale Sur-les-Hautes Pressions, LeCreusot, Saone-et-Loire, France (1965)

19. Forman, R.A., Piermarini, G.J., Barnett, J.D., Block, S.: Science **176**, 284 (1972)
20. Mao, H.K., Bell, P.M.: Science **200**, 1145 (1978)
21. Herber, R.H., Spijkerman, J.: J. Chem. Phys. **42**, 4312 (1965)
22. Huggins, F.E., Mao, H.J., Virgo, D.: Carnegie Institution of Washington Year Book **74**, 405 (1975)
23. Cort, G., Taylor, R.D., Willis, J.O.: J. Appl. Phys. **53**, 2064 (1982)
24. Farrell, J.N.: PhD thesis, University of North Carolina (1984)
25. Chow, L., Dean, P.A., Farrell, J.N., Magill, P.A., Roberts, L.D.: Phys. Rev. B **33**, 3039 (1986)
26. Pasternak, M., Farrell, J.N., Taylor, R.D.: Hyperfine Interact. **28**, 837 (1985)
27. Nasu, S., Kurimoto, K., Nagatomo, S., Endo, S., Fujita, F.E.: Hyperfine Interact. **29**, 1583 (1985)
28. Taylor, R.D., Farrell, J.N.: J. Appl. Phys. **61**, 3669 (1987)
29. Pasternak, M.P., Rozenberg, G.Kh., Machavariani, G.Yu., Naaman, O., Taylor, R.D., Jeanloz, R.: Phys. Rev. Lett. **82**, 4663 (1999)
30. van der Woude, F.: Phys. Status Solidi **17**, 417 (1966)
31. McQueen, R.G., Marsh, S.P.: In: Clark, S.P. (ed.) Handbook in Physical Constants, p. 153. Memoir 97 of the Geological Society of America, Inc., Revised Edition (1966)
32. Reid, A.F., Ringwood, A.E.: J. Geophys. Res. **74**, 3238 (1969)
33. Yagi, T., Akimoto, S.: In: Akimoto, S., Manghnani, M.H. (ed.) High Pressure Research in Geophysics, p. 81. Center Acad. Publ. Japan, Tokyo (1982)
34. Suzuki, T., Yagi, T., Akimoto, A., Ito, A., Morimoto, S., Syono, S.: In: Minomura, S. (ed.) Solid State Physics Under Pressure, p. 149. Terra Scientific (1985)
35. Shannon, R.D., Prewitt, C.T.: J. Solid State Chem. **2**, 134 (1970)
36. Staun Olsen, J., Cousins, C.S.G., Gerward, L., Jhans, H., Sheldon, B.J.: Phys. Scr. **43**, 327 (1991). (These authors were not aware, at this time, of the results of Piermarini and Block, 1975, private communication)
37. X-ray diffraction studies were carried out at the High-Pressure ID30 beam-line at the European Synchrotron Research Facility, Grenoble
38. Machavariani, G.Yu., Pasternak, M.P., Hearne, G.R., Rozenberg, G.Kh.: Rev. Sci. Instrum. **69**, 1423 (1998)
39. The hematite sample for MS studies was enriched to 15% ^{57}Fe
40. Hearne, G.R., Pasternak, M.P., Taylor, R.D.: Rev. Sci. Instrum. **65**, 3787 (1994) (The particular 57Fe MS setup used for high pressure studies with DAC's that used custom-made 0.5 mm×0.5 mm 57Co(Rh) point sources is described.)
41. The ground state $^2T_{2g}$ of the low-spin 5*d*-electrons configuration in Fe^{3+} is $\left(t^*_{2g}\uparrow\right)^3\left(t^*_{2g}\downarrow\right)^2$, with magnetic moment approximately 1/5 of that of the high-spin $^6A_{1g}$ ground state.
42. Note that following the first compression cycle (filled circle curve), $R(P)$ upon decompression reaches a value that is lower at ~5 GPa. Successive compression and decompression cycles are reproducible. This phenomenon can be explained as due to compacting of the sample during the first compression cycle.
43. Pasternak, M.P., Taylor, R.D., Jeanloz, R., Li, X., Nguyen, J.H., McCammon, C.A.: Phys. Rev. Lett. **79**, 5046 (1997)
44. Hubbard, J.: Proc. R. Soc. **A277**, 237 (1964)
45. Pasternak, M.P., Taylor, R.D., Chen, A., Meade, C., Falicov, L.M., Giesekus, A., Jeanloz, R., Yu, P.Y.: Phys. Rev. Lett. **65**, 790 (1990)
46. Pasternak, M.P., Taylor, R.D., Jeanloz R.: In: Hochheimer, H.D., Etters, R.D. (ed.) Frontiers of High Pressure Research, p. 227. Plenum, New York (1992)
47. Hearne, G.R., Pasternak, M.P., Taylor, R.D.: Hyperfine Interact. **90**, 447 (1994)
48. Bargeron, C.B., Avinor, M., Drickamer, H.G.: Inorg. Chem. **7**, 1338 (1971). (By substituting 2% 57Fe in MnS_2 (TN=48 K), claimed to observe a HS→LS transition in Fe impurities starting at ~7 GPa and reaching a complete conversion to the LS state at 12 GPa. MS studies were done at 300 K. The signature for this transition was the different QS and IS claimed to be associated with LS Fe^{2+}.)
49. McCammon, C.A.: J. Magn. Magn. Mater. **104**, 1937 (1992)
50. Pasternak, M.P., Milner, A.P., Rozenberg, G.Kh., Taylor^2, R.D., Jeanloz, R.: Phys. Rev. Lett. **92**, 085506 (2004)
51. Kruger, M.B. Williams, Q., Jeanloz, R.: J. Chem. Phys. **91**, 5910 (1989)
52. Duffy, T.S. Meade, C., Fei, Y., Mao, H.K. Hemley, R.J.: Am. Mineral. **80**, 222 (1995)
53. Nguyen, J.H. Kruger, M.B. Jeanloz, R.: Phys. Rev. Lett. **78**, 1936 (1997)
54. Parise, J.B., Loveday, J.S., Nelmes, R.J., Kagi, H.: Phys. Rev. Lett. **83**, 328 (1999)
55. Miyamoto, H., Shinjo, T., Bando, Y., Takada, T.: J. Phys. Soc. Jpn. **23**, 1421 (1967) ($Fe(OH)_2$ can be classified as a wide gap Mott insulator. At ambient pressure it orders antiferromagnetically with space group $P_{2c}\overline{1}$ at T_N=34 K with the Fe ferromagnetic sublattice spin direction alternating between 90 and 270° with respect to *c*-axis.)
56. Parise, J.B., Marshall, W.G., Smith, R.I., Lutz, H.D., Möller, H.: Am. Mineral. **85**, 189 (2000)

57. Pasternak, M.P., Xu, W.M., Rozenberg, G.Kh., Taylor, R.D., Jeanloz, R.: JMMM **265**, L107 (2003)
58. Verwey, E.J.: Nature (Lond.) **144**, 327 (1939)
59. Verwey, E.J., Haayman, P.W., Romeijan, F.C.: J. Chem. Phys. **15**, 181 (1947)
60. Reviews on several aspects of the Verwey transition published before 1980 are collected in the special issue of Philos. Mag. B 42 no. 3 (1980)
61. Mott, N.F.: Festkorperprobleme **19**, p 331 (1979) (Mott proposed that the electron assembly in magnetite above T_v may be characterized as a Wigner glass; electrons are in a localized state generated through interactions with other localized electrons or with impurities or defects. At $T=T_V$ the Wigner glass transforms discontinuously into a Wigner crystal at lower T, so long as the electron assembly in the coordinated sites is subject to long range order.)
62. Novák, P., Štěpánková, H., Englich, J., Kohout, J., Brabers, V.A.M.: Phys. Rev. B **61**, 1256 (2000)
63. García, J., Subías, G., Proietti, M.G., Blasco, J., Renevier, H., Hodeau, J.L., Joly, Y.: Phys. Rev. B **63**, 054110 (2001)
64. Berry, F.J., Skinner, S., Thomas, M.F.: J. Phys. Condens. Matter **10**, 215 (1998) and references therein
65. Iizumi, M., Koetzle, T.F., Shirane, G., Chikazumi, S., Matsui, M., Todo, S.: Acta Crystallogr. **B38**, 2121 (1982)
66. Nasu, S.: Hyperfine Interact. **90**, 59 (1975)
67. Rozenberg, G.Kh., Milner, A.P., Pasternak, M.P., Hearne, G.R., Taylor, R.D.: Phys. Rev. B **58**, 10283 (1998)
68. Mizokawa, T., Namatame, H., Fujimori, A., Akeyama, K., Kondoh, H., Kuroda, H., Kosugi, N.: Phys. Rev. Lett. **67**, 1638 (1991)
69. Produced by D'Anvils Ltd (http://www.danvils.com)

Hyperfine Interact (2006) 170:33–46
DOI 10.1007/s10751-006-9476-1

Vibrational dynamics of Fe in amorphous Fe–Sc and Fe–Al alloy thin films

B. Sahoo · W. Keune · E. Schuster · W. Sturhahn · J. Zhao · T. S. Toellner · E. E. Alp

Published online: 5 December 2006

Abstract The atomic vibrational dynamics of ^{57}Fe in 800-Å thick amorphous (a-) $^{57}Fe_{0.25}Sc_{0.75}$, a-$^{57}Fe_{0.67}Sc_{0.33}$ and a-$^{57}Fe_{0.14}Al_{0.86}$ alloy thin films has been investigated at room temperature by nuclear resonant inelastic X-ray scattering (NRIXS) of synchrotron radiation. The amorphous phase has been successfully stabilized by codeposition of Fe and Sc or Al in ultrahigh vacuum onto substrates held at −140 °C during deposition. The amorphous structure of the samples was confirmed by X-ray diffraction and conversion electron Mössbauer spectroscopy. The ^{57}Fe-projected partial vibrational density of states, g(E), has been obtained from the measured NRIXS vibrational excitation probability, together with thermodynamic quantities such as the probability of recoilless absorption (f-factor), the average kinetic energy per Fe atom, the average force constant, and the vibrational entropy per Fe atom. A plot of the reduced density of states, g(E)/E^2, versus excitation energy E proves the existence of non-Debye-like vibrational modes (boson peak) with a peak energy, E_{bp}, in the range of 3–7 meV. Both, the boson peak height H_{bp} and E_{bp} were found to depend on the composition. Above the boson peak, g(E)/E^2 exhibits an exponential decrease. Our results demonstrates that the features of the boson peak depend on the amount and type of element M (M = Al, Si, Mg, Sc).

Key words amorphous alloys · Fe–Sc · Fe–Al · thin films · conversion electron Mössbauer spectroscopy (CEMS) · nuclear resonant inelastic X-ray scattering · atomic vibrational density of states · boson peak

B. Sahoo (✉) · W. Keune · E. Schuster
Department of Physics, University of Duisburg-Essen, D-47048 Duisburg, Germany
e-mail: balaram_sahoo@uni-duisburg.de

W. Sturhahn · J. Zhao · T. S. Toellner · E. E. Alp
Advanced Photon Source, Argonne National Laboratory, Argonne, IL 60439, USA

1 Introduction

The atomic vibrational dynamics of amorphous and disordered systems is a subject of continuing interest because of their anomalous behavior in the low energy part of the vibrational density of states (VDOS), g(E) [1, 2]. In particular, the phenomenon of the so called 'boson peak' which appears as an excess contribution at low energies in the reduced VDOS, $g(E)/E^2$, as compared to the usual Debye behavior ($g(E)/E^2$ = constant) of crystalline materials, is a topical subject [3]. The microscopic origin of the boson peak is still a matter of debate [3–7]. The existence of such excess vibrational excitations in different bulk materials [8] and in particular in metallic glasses [9] was revealed by earlier inelastic neutron scattering results. More recently, nuclear resonant inelastic X-ray scattering (NRIXS) of synchrotron radiation at ^{57}Fe nuclei has been used to measure the boson peak via ferrocene probe molecules matrix-isolated in various bulk organic glasses [3], and in various amorphous binary alloy thin-film systems [10–18]. A systematic NRIXS investigation of vapor-quenched amorphous (a-) Fe_xTb_{1-x} [16] and a-Fe_xMg_{1-x} [18] alloy thin films was reported recently. These results demonstrate that the features of the boson peak in a-Fe_xM_{1-x} alloys depends on the alloy composition and on the type of alloying element M. Therefore, in order to obtain information on the nature of the boson peak in metallic glasses it is essential to study systematically the vibrational properties of binary a-Fe_xM_{1-x} alloys as a function of atomic mass, atomic volume and electronegativity of the element M. The present work is a continuation of our previous systematic studies of the properties of the boson peak in amorphous binary-alloy thin films. Here, we present results on a-Fe–Sc and a-Fe–Al alloy thin films, and we compare with properties of the a-Fe–Mg system.

For our investigation we have employed ^{57}Fe NRIXS. After the pioneering work by different groups [19–21] the NRIXS method has been widely applied to explore the atomic vibrational properties of condensed matter, see, e.g., ref. [22–25]. The frequency (or energy) distribution of atomic vibrations, g(E), for the vibrating resonant isotope in the crystalline or amorphous state, can be directly measured by NRIXS. The VDOS is a fundamental quantity from which important thermodynamic properties may be deduced [26–29]. Because only small quantities of materials are needed, NRIXS has opened the field of lattice dynamics in thin film, interface and multilayer research [10–18, 30].

2 Experimental procedure and sample preparation

Thin films of a-$^{57}Fe_{0.25}Sc_{0.75}$, a-$^{57}Fe_{0.67}Sc_{0.33}$ and a-$^{57}Fe_{0.14}Al_{0.86}$ alloys have been prepared in ultrahigh vacuum (UHV) by thermal co-evaporation of metallic ^{57}Fe (95 % isotopic enrichment) and Sc or Al from Knudsen cells (Al_2O_3 crucibles) onto oxidized Si(001) substrates held at −140 °C during deposition. The base pressure was about 5×10^{-10} mbar. The pressure during evaporation was $\sim 2 - 3 \times 10^{-9}$ mbar. The deposition rates of ^{57}Fe and Sc or Al were measured independently by two calibrated quartz-crystal oscillators and controlled by a personal computer. The desired alloy composition was obtained by selecting a constant ratio of ^{57}Fe and Sc (or Al) deposition rates, while maintaining the individual deposition rates constant. The individual deposition rate was at a constant value between 0.1 and 0.4 Å/s.

The alloy compositions, x, given here are nominal compositions determined from the ratio of the two deposition rates. The error in x is estimated to be $\pm$ 0.01. The thickness of the alloy films was 800 Å as monitored by the quartz crystals. The purity of the elements used was 99.95% for ^{57}Fe, 99.9% for Sc and 99.9999% for Al. The alloy films were coated by a 60-Å thick Cr layer for protection.

The structure of the samples was characterized by conventional $\theta - 2\theta$ X-ray diffraction (XRD) and ^{57}Fe conversion electron Mössbauer spectroscopy (CEMS). For the XRD measurements Cu–K_α radiation and a graphite monochromator were used. The CEM spectra were measured at room temperature by placing the sample into a He + 4% CH_4 proportional counter with the film surface perpendicular to the incident 14.4 keV γ-ray of the ^{57}Co source (Rh-matrix). The spectra were least-squares fitted by a distribution of quadrupole doublets. For the least-squares fitting of the CEM spectra, the computer program *NORMOS* by Brand [31] was used.

The ^{57}Fe NRIXS experiments were performed at RT at the undulator beamline 3-ID of the Advanced Photon Source in Argonne (USA). The method of inelastic nuclear resonant absorption of 14.4125 keV X-rays is selective to the ^{57}Fe resonant isotope and provides the Fe-projected (partial) VDOS rather directly with a minimum of modeling [20, 27]. Details of the technique are described elsewhere [19–21, 26–29, 32]. The monochromatized synchrotron radiation was incident onto the film surface under a grazing angle of $\sim$ 4 mrad and had an energy bandwidth (FWHM) of 1 meV. The X-ray beam was focussed to about $10 \times 10\ \mu m^2$ in size. The energy was tuned around the 14.4125 keV nuclear resonance of ^{57}Fe. Avalanche photo diodes were used as detectors. The NRIXS detector was placed close to the sample surface in order to detect the characteristic K-fluorescence X-rays emitted by Fe atoms during the process of de-excitation of ^{57}Fe nuclei to the ground state. The instrumental resolution function was measured simultaneously with each NRIX spectrum by detecting the forward scattered intensity. The measurement time per spectrum was about 1 h. The NRIXS data evaluation and extraction of the VDOS were performed by using the computer program PHOENIX described elsewhere [33]. The thickness of 800 Å of our films is sufficiently large, so that their measured VDOS distributions are representative of the bulk materials [11, 14, 16].

3 Sample characterization

Figure 1 shows typical XRD patterns of our samples. Only sharp Bragg reflections from the Si substrate and a weak Bragg peak from the Cr overlayer are observed in Figure 1, but no sharp Bragg reflections from the alloy films. This demonstrates the amorphous nature of our alloy films. There is no indication of precipitation of any crystalline phase. The amorphous structure of the alloy films is revealed by faint and broad features in the region around $2\Theta = 38° \sim 44°$ for Fe–Sc and $2\Theta = 35° \sim 44°$ for Fe–Al, where the intensity is weakly enhanced and behaves non-monotonically relative to the decaying (extrapolated) background intensity. These weak and broad features are indicative of a broad distribution of local short-range-ordered atomic arrangements (but not of long-range order) in the amorphous alloy films.

The CEM spectra (Figure 2) display apparent quadrupole doublets with broadened lines for all samples. Similar asymmetric doublets are typical for non-magnetic binary amorphous alloys [34–38]. The change in the apparent line intensity ratio

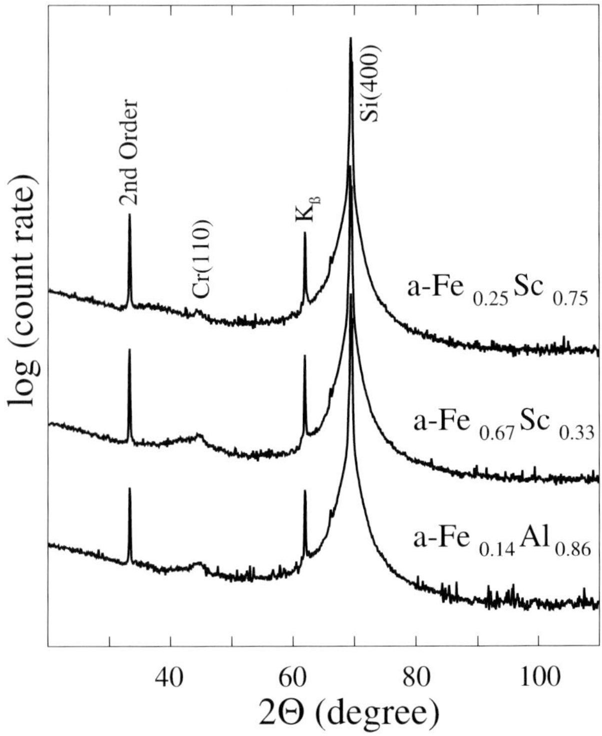

Figure 1 Θ-2Θ X-ray diffraction pattern of 800 Å thick a-$Fe_{0.25}Sc_{0.75}$, a-$Fe_{0.67}Sc_{0.33}$ and a-$Fe_{0.14}Al_{0.86}$ alloy thin films grown on oxidized Si(100) substrates at $T_s = -140$ °C. Cu–K_α radiation was used. K_β means: Si(400)–Cu–K_β peak, and 2nd order means: Si(400)–Cu–K_α, 2nd order reflection. Cr(110): reflection from Cr coating layer. Si(400): reflection from Si(100) substrate.

observed in Figure 2a as compared to Figure 2b for Fe–Sc indicates a modification of the local atomic environment around the ^{57}Fe nuclei with increasing Fe content in the amorphous structure. It is interesting that the reported spectral asymmetry for a-Fe_xMg_{1-x} alloy films [18] is opposite to that observed here for the a-Fe_xSc_{1-x} system, i.e. Fe-rich amorphous Fe–Mg alloy films are found to have a similar intensity asymmetry as Fe-poor Fe–Sc amorphous films. This means that at the same composition the atomic short-range order in amorphous Fe–Sc and Fe–Mg alloys is different .

All CEM spectra in Figure 2 were least-squares fitted with a distribution P(QS) of quadrupole splittings (QS), as shown on the right-hand side of the corresponding spectrum in Figure 2. Good fits were obtained by assuming a linear correlation between QS (identical to ΔE_Q) and the isomer shift δ, which often occurs in binary disordered systems [34–38]. The obtained Mössbauer spectral parameters are listed in Table I. Negative isomer shift values have been reported also in bulk a-$FeSc_3$ alloys [36]. For a-$Fe_{0.14}Al_{0.86}$, the average quadrupole splitting <QS> obtained here at RT (<QS> = 0.48 mm/s) is somewhat smaller than the value <QS> = 0.6 mm/s reported for bulk a-$FeSc_3$ at 4.2 K [36]. This could be due to the different measurement temperature and/or differences in the atomic short-range order of thin films as compared to bulk samples. Our values of <QS> = 0.48 mm/s for a-$Fe_{0.25}Sc_{0.75}$ is not very different from that found for amorphous Zr-rich bulk-ZrFe alloys (<QS> = 0.52 mm/s, [37]). For our Fe-rich a-$Fe_{0.67}Sc_{0.33}$ film at RT, the average isomer shift ($\langle\delta\rangle = -0.064$ mm/s) and average quadrupole splitting

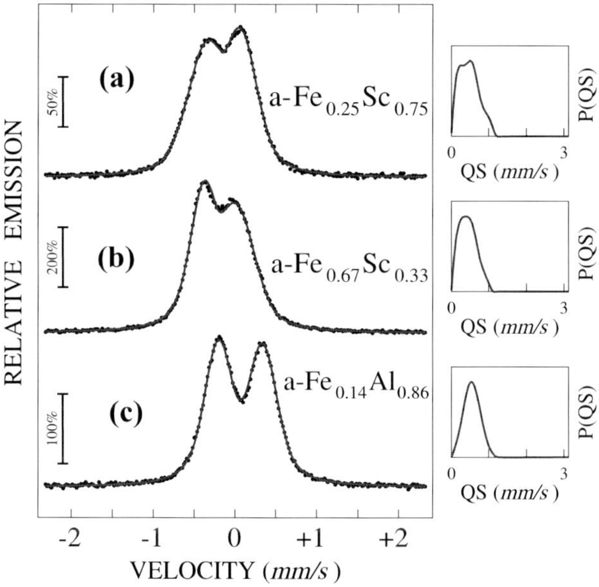

Figure 2 RT Mössbauer spectra (CEMS) of 800-Å thick a-$Fe_{0.25}Sc_{0.75}$ (**a**), a-$Fe_{0.67}Sc_{0.33}$ (**b**) and a-$Fe_{0.14}Al_{0.86}$ (**c**) alloy thin films. The spectra were least-squares fitted by a distribution P(QS) of quadrupole splittings (QS), as shown on the right-hand side.

Table I Mössbauer spectral parameters

Sample	$<\delta>$	<QS>
a-$Fe_{0.25}Sc_{0.75}$	−0.039	0.48
a-$Fe_{0.67}Sc_{0.33}$	−0.064	0.45
a-$Fe_{0.14}Al_{0.86}$	0.176	0.56

$<\delta>$ = average isomer shift (in mm/s, relative to bulk α-Fe at room temperature), <QS> = average quadrupole splitting (= ΔE_Q, in mm/s) at room temperature

(<QS> = 0.45 mm/s) are in reasonable agreement with corresponding 300-K values of (non-magnetic) Fe-rich bulk a-$Fe_{0.90}Sc_{0.10}$ alloys ($<\delta>= -0.09$ mm/s, <QS> = 0.40 mm/s [38]). The quadrupole splitting <QS> of a-$^{57}Fe_{0.14}Al_{0.86}$ (Table I) is much larger than the <QS>-values reported for crystalline Al-rich Fe–Al phases, such as $FeAl_6$ (0.26 mm/s), Fe_4Al_{13} (0 and 0.40 mm/s, respectively), and Fe_2Al_5 (0.46 mm/s) [39]. As a conclusion, we have demonstrated that our quench-condensed Fe–Sc and Fe–Al alloy films are amorphous.

4 NRIXS results and discussion

4.1 g(E) and related quantities

Typical NRIXS spectra (raw data) of a-Fe_xSc_{1-x} ($x = 0.25$ and 0.67) and a-$Fe_{0.14}Al_{0.86}$ alloy thin films are shown in Figure 3a,b and Figure 4a, respectively. These spectra which give the nuclear excitation probability versus energy transfer E, show the dominant central elastic peak at the nuclear transition energy E_0 (energy transfer E = 0 meV) and sidebands at lower and higher energy. The low-energy sideband is

Figure 3 NRIXS spectra (raw data) of 800-Å thick a-$Fe_{0.25}Sc_{0.75}$ (**a**) and a-$Fe_{0.67}Sc_{0.33}$ (**b**) alloy thin films grown at $T_s = -140$ °C. The instrumental resolution function (central elastic peak, dotted-hatched curve) is also shown. (For clarity the spectra are vertically displaced). The spectra were taken at room temperature. Vibrational excitation probability per unit energy, W(E), for a-$Fe_{0.25}Sc_{0.75}$ (**c**) and a-$Fe_{0.67}Sc_{0.33}$ (**d**) samples at room temperature, obtained from (**a**) and (**b**), respectively, after subtraction of the central elastic peak and proper normalization. The contribution of one-phonon, two-phonon and higher-order phonons are also shown.

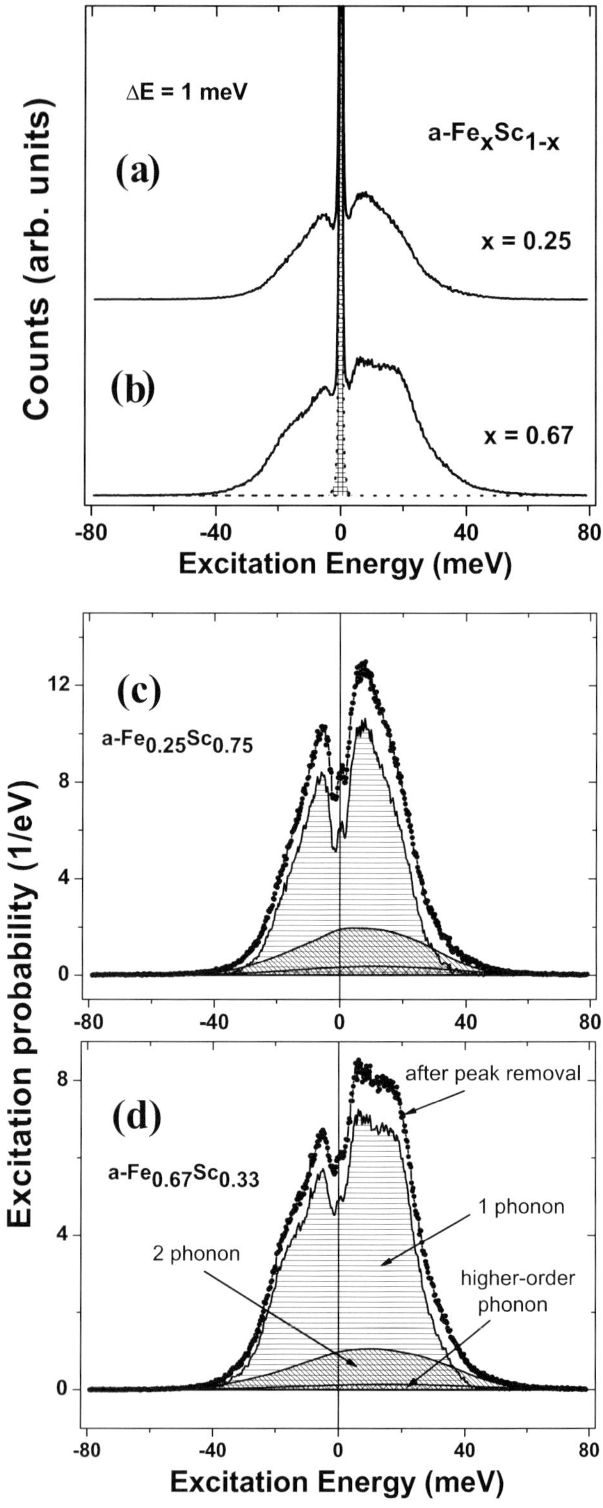

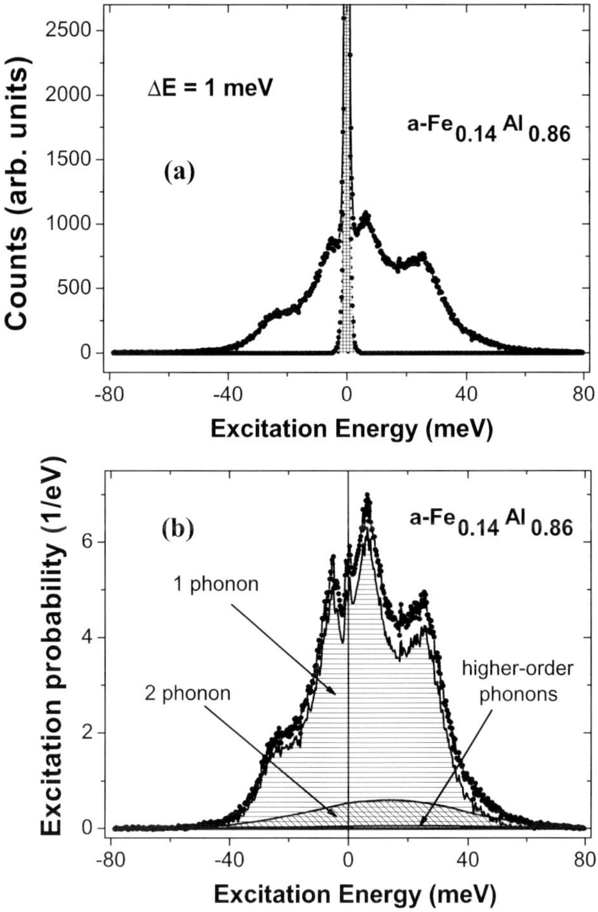

Figure 4 (**a**) NRIXS spectra (raw data) of 800-Å thick a-$Fe_{0.14}Al_{0.86}$ alloy thin film grown at $T_s = -140\ °C$. The instrumental resolution function (central elastic peak, dotted-hatched curve) is also shown. The spectrum was taken at room temperature. (**b**) Vibrational excitation probability per unit energy, W(E), for a-$Fe_{0.14}Al_{0.86}$ sample at room temperature, obtained from **a** after subtraction of the central elastic peak and proper normalization. The contribution of one-phonon, two-phonon and higher-order phonons is also shown.

produced by net annihilation of vibrational quanta, while the high-energy sideband is the result of creation of vibrational quanta. The observed asymmetry reflects the "detailed balance" due to the Boltzmann factor [40]. The measured instrumental resolution function with a FWHM of 1 meV is also shown in Figure 3a,b and Figure 4a. The resolution function is nearly symmetrical and falls off rapidly from the center without having extended wings. The spectra in Figure 3a,b for a-Fe–Sc do not show sharp features which one encounters for crystalline samples [16, 17]. This is indicative of the amorphous nature of the a-Fe_xSc_{1-x} thin films. On the other hand, somewhat sharper features can be observed in Figure 4a for a-$Fe_{0.14}Al_{0.86}$, which demonstrates that the degree of amorphicity in the a-Fe–Al film is less than that in the a-Fe–Sc films. Apparently, a higher degree of some atomic short-range order exists in the a-$Fe_{0.14}Al_{0.86}$ film than in the a-Fe_xSc_{1-x} films. Obviously NRIXS spectra are a useful qualitative probe for the degree of amorphicity in the sample.

After subtraction of the central elastic peak from the measured NRIXS spectra the data were normalized according to the standard procedures [20, 27, 32] yielding the vibrational excitation probability per unit energy, W(E), as shown in Figure 3c,d for a-Fe_xSc_{1-x} and in Figure 4b for a-$Fe_{0.14}Al_{0.86}$. The contributions to W(E) of

Table II Thermodynamic properties of a-$Fe_{0.25}Sc_{0.75}$, a-$Fe_{0.67}Sc_{0.33}$ and a-$Fe_{0.14}Al_{0.86}$ alloy thin films derived from the NRIXS results: f = f-factor at RT, f_0 = f-factor at 0 K (calculated)

	a-$Fe_{0.25}Sc_{0.75}$	a-$Fe_{0.67}Sc_{0.33}$	a-$Fe_{0.14}Al_{0.86}$
f	0.5844(5)	0.6531(4)	0.7242(4)
f_0	0.8893(3)	0.9002(1)	0.9156(3)
$< x^2 >$	0.01007	0.00799	0.00605
E_D	23.77(2)	26.78(2)	30.92(4)
Θ_D	276	310	359
$V(\vec{s})$	135(4)	127(1)	179(3)
$T(\vec{s})$	13.88(9)	13.84(4)	14.21(7)
$T_0(\vec{s})$	5.43(8)	5.63(3)	6.76(5)
C_v	2.79(1)	2.796(6)	2.71(1)
S_v	4.01(1)	3.767(6)	3.25(9)

The mean square displacement $< x^2 >$ (in Å^2), the Debye energy E_D (in meV) and the Debye temperature Θ_D (in K) were calculated from f at RT. The values at T = 0 K were calculated from corresponding RT values

inelastic processes in terms of one-phonon, two-phonon and higher-order phonons are also depicted in Figure 3c,d and Figure 4b. They were obtained according to the procedures described in refs. [20, 27, 32]. If we compare the Fe-poor cases of a-$Fe_{0.25}Sc_{0.75}$ and a-$Fe_{0.14}Al_{0.86}$, one can notice from Figure 3c and Figure 4b that two-phonon and higher-order phonons contribute to a higher extent in a-$Fe_{0.25}Sc_{0.75}$ than in a-$Fe_{0.14}Al_{0.86}$.

The integration of W(E) over all energies E provides the quantity (1 - f) in a model independent way with high precision, where f is the Lamb–Mössbauer factor (or f-factor). Values of the f-factors at RT, obtained from Figure 3c,d and Figure 4b, respectively, are given in Table II. The f-value for the a-Fe–Al case is found to be significantly higher than for the a-Fe–Sc case.

In the harmonic approximation the Fe-projected partial VDOS, g(E), is proportional to the one-phonon contribution $S_1(E)$ in W(E), and is given by [29, 32]

$$g(E) = \frac{S_1(E)}{E_R} E(1 - e^{E/k_B T}), \tag{1}$$

where E_R is the recoil energy of the resonating nucleus, k_B is Boltzmann's constant, and E is the transferred energy. g(E) was deduced from the measured distribution W(E) in Figure 3c,d and Figure 4b according to standard procedures described elsewhere [20, 27, 33].

Figure 5a exhibits g(E) of our amorphous Fe–Sc and Fe–Al alloy films, along with the VDOS for bulk bcc-Fe for comparison. The energy resolution was 1 meV for all g(E) curves. In contrast to the case of crystalline phases (e.g. bcc-Fe) all the VDOS of our samples have the common feature that they do not show sharp peaks, since such peaks (e.g. due to van Hove singularities) are smeared out in structurally and atomically disordered materials. This feature is similar to that of the VDOS of a-Fe_xTb_{1-x} [14, 16] or a-Fe_xMg_{1-x} [18] alloy thin films and bulk a-$Mg_{0.7}Zn_{0.3}$ metallic glass [9].

There is a noticeable difference in the shape of g(E) between the VDOS of the a-Fe-Sc and a-Fe–Al alloy films (Figure 5a). The g(E) curves of a-Fe–Sc are asymmetric and almost structureless broad features: with increasing E the g(E) curves

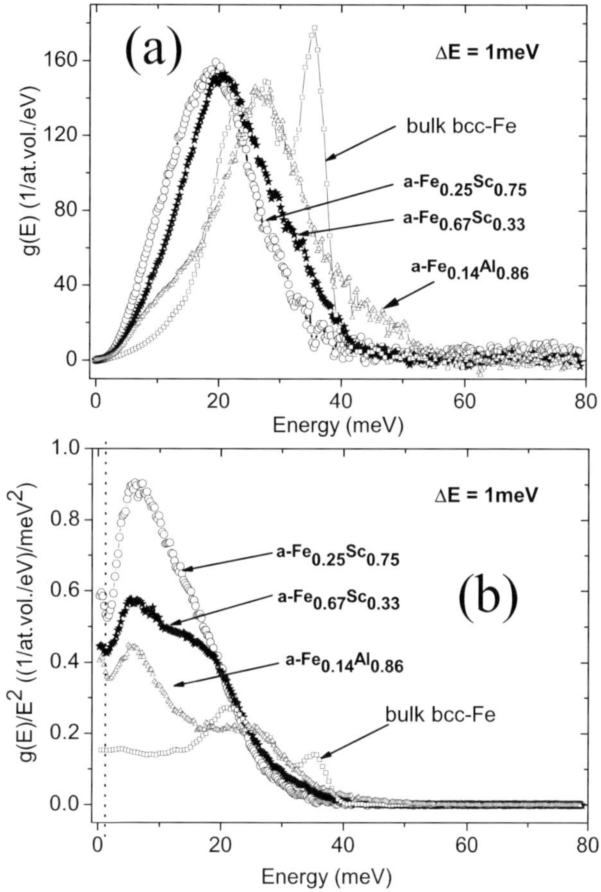

Figure 5 (**a**) Partial VDOS, g(E), of 800 Å thick a-$Fe_{0.25}Sc_{0.75}$, a-$Fe_{0.67}Sc_{0.33}$ and a-$Fe_{0.14}Al_{0.86}$, alloy thin films grown at −140 °C. The VDOS of bcc-Fe is also shown for comparison. (**b**) Reduced partial VDOS, g(E)/E^2, versus excitation energy E for the a-$Fe_{0.25}Sc_{0.75}$, a-$Fe_{0.67}Sc_{0.33}$ and a-$Fe_{0.14}Al_{0.86}$ samples at room temperature. Energy resolution $\Delta E = 1.0$ meV. The vertical dotted line indicates the FWHM of 1.0 meV of the instrumental resolution function (data below 1 meV are physically irrelevant). For comparison the reduced VDOS of bulk bcc-Fe is also shown.

first rise steeply and nearly linearly, show a maximum at ∼19 meV for $x = 0.25$ and at ∼21 meV for $x = 0.67$, and then drop more slowly to zero. By contrast, g(E) of a-$Fe_{0.14}Al_{0.86}$ first shows a distinct shoulder near ∼10 meV, a maximum near ∼26 meV, and again a broad shoulder centered near ∼40 meV (not far away from the 36-meV peak of bulk bcc-Fe). These low-energy and high-energy shoulders in g(E) for a-$Fe_{0.14}Al_{0.86}$ could originate from the higher degree of atomic short range order in this alloy, leading possibly to Fe nanoclusters. This effect is much less pronounced or negligible in the amorphous Fe–Sc alloy films, which obviously have a much smaller degree of short range order than the a-$Fe_{0.14}Al_{0.86}$ film. Similar effects have been observed in the VDOS of a-Fe_xMg_{1-x} alloy films and were tentatively assigned to statistical Fe clusters in the amorphous alloy [18].

From the measured W(E) and g(E) spectra important thermodynamic quantities are obtained [29, 33]. Besides the f-factor, we have determined the mean atomic force constant $V(\vec{s})$ along the beam direction $\vec{s}$, the mean kinetic energy per atom $T(\vec{s})$, the vibrational specific heat per atom C_v, and the vibrational entropy per atom S_v, all at RT. These quantities are given in Table II, together with the calculated values of the mean square displacement $< x^2 >$ of the ^{57}Fe atom, the Debye energy E_D and the Debye temperature Θ_D. The values at T = 0 K for the f-factor, f_0, and the mean kinetic energy, $T_0(\vec{s})$, were calculated from the corresponding RT values by using the Debye model.

4.2 The boson peak

The low-energy modes in g(E) below ~ 10 meV in Figure 5a are of particular interest. The usual way to observe a deviation from Debye-like behavior ($g(E) \propto E^2$) is to plot the reduced VDOS, $g(E)/E^2$, versus E. Then, the strict Debye-like behavior is reflected by a horizontal line which intersects the $g(E)/E^2$ axis at a value that is proportional to c_s^{-3} (c_s = average sound velocity) [27, 28].

Figure 5b shows a plot of $g(E)/E^2$ versus E, deduced from Figure 5a. For decreasing E, a remarkable and unambiguous rise in $g(E)/E^2$ and a peak near $E_{bp} \approx 5 \sim 6$ meV are observed for all amorphous alloy films. For comparison, Debye-like behavior of bulk bcc-Fe is observed in $g(E)/E^2$ below ~ 15 meV, where $g(E)/E^2$ remains constant at ~0.15 $(\text{at.vol})^{-1}$ $(\text{eV})^{-1}$ $(\text{meV})^{-2}$ [14] as shown in Figure 5b. The strong rise and the peak in $g(E)/E^2$ for the present amorphous alloy films is qualitatively similar to the $g(E)/E^2$-behavior in a-Fe_xTb_{1-x} [14, 16] and a-Fe_xMg_{1-x} [18] alloy thin films. The peak height, H_{bp}, is observed to increase with decreasing Fe content. Non-Debye-like low-energy vibrational excitations are clearly present in the a-Fe_xSc_{1-x} and a-$Fe_{0.14}Al_{0.86}$ alloy thin films. The observed peak in Figure 5b is identified as the boson peak.

The probability of excitation of these anomalous low-energy atomic vibrations increases with decreasing Fe content x. This is displayed in Figure 6a, where we show the concentration dependence of the boson peak height, H_{bp}, for our a-Fe_xSc_{1-x} alloy films and for comparison, for a-Fe_xMg_{1-x} alloy films reported earlier [18]. Data points for a-$Fe_{0.14}Al_{0.86}$ and for a-$FeSi_2$ [17] are also given. (H_{bp} was measured from $g(E)/E^2 = 0$ in Figure 5b). For a-Fe_xMg_{1-x} H_{bp} increases linearly with composition $(1 - x)$ with a slope S_{FeMg} equal to 1.17 $(\text{at. vol.})^{-1}(\text{eV})^{-1}(\text{meV})^{-2}$. A linear behavior was observed also for a-Fe_xTb_{1-x} alloy thin films [14, 16]. Therefore, we may assume that a linear relationship exists also for a-Fe_xSc_{1-x}, although we have measured only two data points. For the straight line connecting the data for a-Fe_xSc_{1-x} in Figure 6a we infer a slope S_{FeSc} of 0.78 $(\text{at. vol.})^{-1}(\text{eV})^{-1}(\text{meV})^{-2}$, which is smaller than for the a-Fe_xMg_{1-x} case. The slope S_{FeTb} for a-Fe_xTb_{1-x} alloy films was reported to be much larger, namely 2.25 $(\text{at. vol.})^{-1}(\text{eV})^{-1}(\text{meV})^{-2}$ [16]. This demonstrates that the slope of the linear H_{bp} vs. $(1 - x)$ behavior does not simply depend on the atomic mass m of the alloying element, because $m_{Tb} > m_{Sc} > m_{Mg}$, but $S_{FeTb} > S_{FeMg} > S_{FeSc}$. Very likely also other properties of the alloying elements, as, e.g., atomic volume or electronegativity, might play a role for the description of the boson peak height. In any case, the anomalous low-energy excitations in these amorphous Fe alloys are related to vibrational modes induced by the non-ferrous atoms.

Figure 6b shows the concentration dependence of the boson peak energy, E_{bp}, for a-Fe_xSc_{1-x} and, for comparison, for a-Fe_xMg_{1-x} [18]. For a-Fe_xMg_{1-x}, E_{bp} decreases linearly with composition $(1 - x)$, with a rather large slope of -5.03 meV. By contrast, the slope of the straight line connecting the two data points for a-Fe_xSc_{1-x} is very small, namely only -0.96 meV. At present there are no systematic studies on which factors influence E_{bp} and its concentration dependence in amorphous binary alloys.

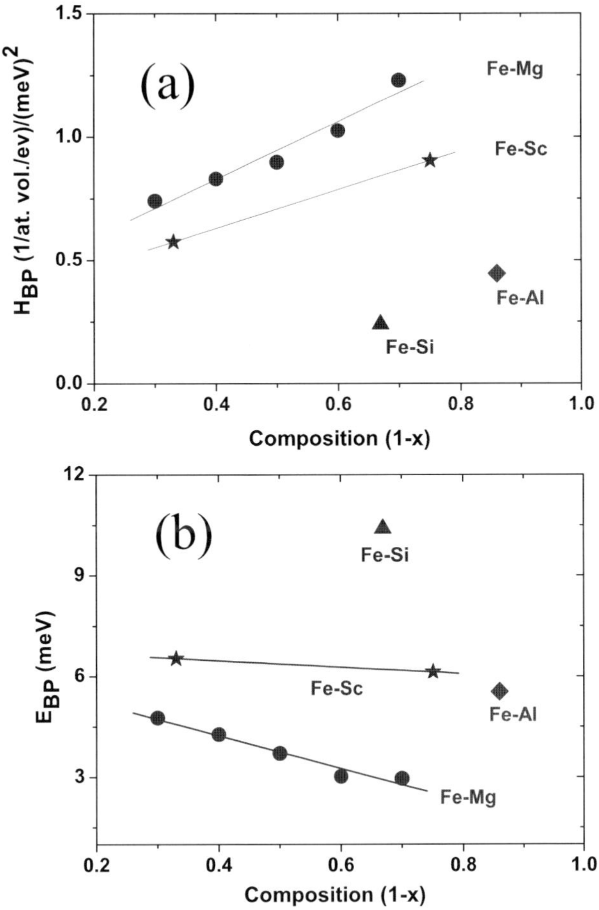

Figure 6 The boson peak height H_{bp} (**a**) and boson peak energy E_{bp} (**b**) plotted versus the composition (1-x) of a-Fe_xM_{1-x} alloy thin films (M = Sc, Al, Si [17] and Mg [18]) at room temperature. The straight lines are least-squares fits to the data points.

4.3 Scaling

Based on experiments and calculations Chumakov et al. [3] have demonstrated that at energies above the boson peak $g(E)/E^2$ exhibits a universal exponential decrease, i.e. $g(E)/E^2 \propto exp(-E/E_0)$, with a parameter E_0 close to the boson peak energy, E_{bp}. Figure 7 shows a plot of $\log[g(E)/E^2]$ versus E for our a-Fe_xSc_{1-x} and a-$Fe_{0.14}Al_{0.86}$ alloy films. As can be seen in Figure 7, directly above the boson peak, i.e. in the range ~ 6 meV $\leq E \leq\sim 15$ meV, the experimental data follow approximately an exponential behavior. From straight line fits (Figure 7, inserts), values of E_0 = 37.7, 106.0, and 26.6 meV for a-$Fe_{0.25}Sc_{0.75}$, a-$Fe_{0.67}Sc_{0.33}$ and a-$Fe_{0.14}Al_{0.86}$, respectively, are obtained. These values are larger than $E_{bp} \sim$ 3–5 meV. Further, a second region with an exponential decrease is observed in Figure 7 at higher energies, i.e. in the range of ~ 20 meV $\leq E \leq\sim 30$ meV for a-Fe_xSc_{1-x}, and ~ 27 meV $\leq E \leq\sim 50$ meV for a-$Fe_{0.14}Al_{0.86}$. Straight line fits result in slopes equivalent to E_0 = 11.4, 15.2 and 14.5 meV for a-$Fe_{0.25}Sc_{0.75}$, a-$Fe_{0.67}Sc_{0.33}$ and a-$Fe_{0.14}Al_{0.86}$, respectively. These values are also larger than the boson peak energies, E_{bp}. A similar behavior with two regions of exponential decrease in $g(E)/E^2$ was also observed in a-$^{57}FeSi_2$ [17] and

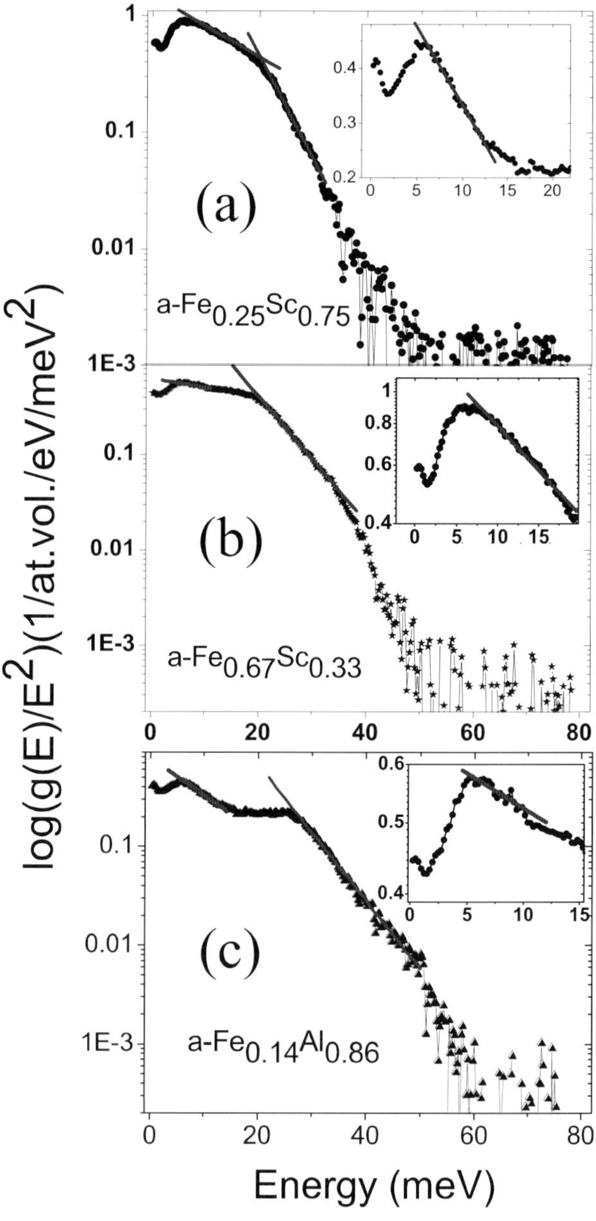

Figure 7 Reduced partial VDOS, $g(E)/E^2$, at room temperature on a logarithmic scale versus energy E for a-$Fe_{0.25}Sc_{0.75}$ (**a**), a-$Fe_{0.67}Sc_{0.33}$ (**b**) and a-$Fe_{0.14}Al_{0.86}$ (**c**) alloy thin films. Inserts: $\log(g(E)/E^2)$ on an expanded energy scale, showing the linear region above the boson peak. The straight lines are least-squares fits to the data.

a-Fe_xMg_{1-x} [18] alloy films. Further investigations on other systems are required in order to prove whether such a behavior is universal.

5 Summary

Vapor-quenched ^{57}Fe-enriched amorphous (a-)Fe_xSc_{1-x} alloy thin films ($x = 0.25$ and 0.67) and a-$Fe_{0.14}Sc_{0.86}$ alloy films have been successfully prepared by co-deposition of the elements. The amorphous structure was confirmed by X-ray diffraction and ^{57}Fe CEMS. The samples were investigated at RT by NRIXS of 14.4125 keV synchrotron radiation with 1 meV energy resolution. The Fe projected VDOS, g(E), and various thermodynamic quantities were deduced from the NRIXS spectra. A plot of $g(E)/E^2$ versus E proves the existence of non-Debye-like vibrational excitations at low-energies, with a peak near $E_{bp} \sim$ 3–5 meV (boson peak). For a-Fe_xSc_{1-x} the boson peak height, H_{bp}, was found to increase with Sc content $(1-x)$, in agreement with the behavior for other amorphous alloy systems, e.g. a-Fe_xMg_{1-x} [18] and a-Fe_xTb_{1-x} [16]. Hence, the anomalous low-energy excitations in these amorphous alloys are related to vibrational modes induced by the non-ferrous atoms. The slope of the linear relationship between H_{bp} and composition $(1-x)$ observed for a-Fe_xMg_{1-x} [18], a-Fe_xTb_{1-x} [16] and (presumably) a-Fe_xSc_{1-x} shows a non-monotonic behavior with respect to the atomic masses of Mg, Tb and Sc. Therefore, in order to describe the composition dependence of H_{bp}, other factors than the atomic mass, as, e.g., the atomic volume and the electronegativity of the non-ferrous elements, could play an important role. At energies E directly above the boson peak $g(E)/E^2$ vs. E shows a narrow region with an exponential decrease, as theoretically predicted and observed by Chumakov et al.[3]. Moreover, much farther above the boson peak energy E_{bp}, a second region with an exponential decrease of $g(E)/E^2$ was found for a-Fe_xSc_{1-x} and a-$Fe_{0.14}Al_{0.86}$, similar to the case of a-Fe_xMg_{1-x} [18] and a-$FeSi_2$ [17] alloy thin films. To the best of our knowledge no theoretical model exists for the description of the latter behavior.

It has been suggested that either vibrations of atoms on the surface of very small voids or oscillations in low density 'defect' regions of metallic glasses are responsible for the anomalous low energy excitations [9]. We suggest that the large atomic radius of the non-ferrous elements (1.65 Å for Sc, 1.43 Å for Al, 1.60 Å for Mg, and 1.79 Å for Tb [41]) as compared to that of Fe (1.24 Å) is an important factor for the understanding of the magnitude of the boson peak. The observed linear increase of the boson peak height, H_{bp}, with the content $(1-x)$ of the non-ferrous element (Figure 6a) could originate from a linear increase of the density of voids with increasing $(1-x)$.

Acknowledgements We are grateful to U. von Hörsten (Duisburg) for valuable technical assistance and sample preparation. Work at Duisburg was supported by Deutsche Forschungsgemeinschaft (GRK 277 and SFB 491). Work at Argonne was supported by the US department of energy, Basic Energy Sciences, Office of Science, under Contract No. W-31-109-Eng-38.

References

1. Phillips, W.A. (ed.): Amorphous Solids – Low Temperature Properties. Springer, Berlin Heidelberg New York (1993)
2. Dianoux, A.J., Petry, W., Richter, D. (eds.): Dynamics of Disordered Materials II. North-Holland, Amsterdam, The Netherlands (1993)
3. Chumakov, A.I., Sergueev, I., van Bürck, U., Schirmacher, W., Asthalter, T., Rüffer, R., Leupold, O., Petry, W.: Phys. Rev. Lett. **92**, 245508 (2004)
4. Finkemeier, F., von Niessen, W.: Phys. Rev., B **66**, 087202 (2002)
5. Schirmacher, W., Diezemann, G., Ganter, C.: Phys. Rev. Lett. **81**, 136 (1998)
6. Schirmacher, W., Diezemann, G., Ganter, C.: Physica, B **284-288**, 1147 (2000)
7. Mauer, E., Schirmacher, W.: J. Low Temp. Phys. **137**, 453 (2004)

8. Malinkovsky, V.K., Novikov, V.N., Parshin, P.P., Sokolov, A.P., Zemlyanov, M.G.: Europhys. Lett. **11**, 43 (1990)
9. Suck, H.B., Rudin, H.: In: Beck, H., Güntherodt, H.J. (eds.) Glassy Metals II Topics in Applied Physics vol. 53, p. 217. Springer, Berlin Heidelberg New York (1983)
10. Sturhahn, W., Röhlsberger, R., Alp, E.E., Ruckert, T., Schrör, H., Keune, W.: J. Magn. Magn. Mater. **198/199**, 590 (1999)
11. Keune, W., Sturhahn, W.: Hyperfine Interact. **123/124**, 847 (1999)
12. Ruckert, T., Keune, W., Sturhahn, W., Hu, M.Y., Sutter, J.P., Toellner, T.S., Alp, E.E.: Hyperfine Interact. **126**, 363 (2000)
13. Roldan Cuenya, B., Keune, W., Sturhahn, W., Toellner, T.S., Hu, M.Y.: Phys. Rev., B **64**, 235321 (2001)
14. Ruckert, T., Keune, W., Sahoo, B., Sturhahn, W., Toellner, T.S., Alp, E.E., Röhlsberger, R.: Hyperfine Interact. **144/145**, 65 (2002)
15. Ruckert, T., Keune, W., Sturhahn, W., Alp, E.E.: J. Magn. Magn. Mater. **240**, 562 (2002)
16. Keune, W., Ruckert, T., Sahoo, B., Sturhahn, W., Toellner, T.S., Alp, E.E., Röhlsberger, R.: J. Phys., Condens. Matter **16**, S379 (2004)
17. Walterfang, M., Keune, W., Schuster, E., Zayak, A.T., Entel, P., Sturhahn, W., Toellner, T.S., Alp, E.E., Jochym, P.T., Parlinski, K.: Phys. Rev., B **71**, 035309 (2005)
18. Sahoo, B., Keune, W., Sturhahn, W., Toellner, T.S., Alp, E.E.: J. Phys. Chem. Solids **66**, 2263 (2005)
19. Seto, M., Yoda, Y., Kikuta, S., Zhang, X.W., Ando, A.: Phys. Rev. Lett. **74**, 3828 (1995)
20. Sturhahn, W., Toellner, T.S., Alp, E.E., Zhang, X.W., Ando, M., Yoda, Y., Kikuta, S., Seto, M., Kimball, C.W., Dabrowski, B.: Phys. Rev. Lett. **74**, 3832 (1995)
21. Chumakov, A.I., Rüffer, R., Grünsteudel, H., Grünsteudel, H.F., Grübel, G., Metge, J., Leupold, O., Goodwin, H.A.: Europhys. Lett. **30**, 427 (1995)
22. Parak, F., Achterhold, K.: Hyperfine Interact. **123/124**, 825 (1999)
23. Grünsteudel, H., Paulsen, H., Winkler, H., Trautwein, A.X., Toftlund, H.: Hyperfine Interact. **123/124**, 841 (1999)
24. Lubbers, R., Grunsteudel, H.F., Chumakov, A.I., Wortmann, G.: Science **287**, 1250 (2000)
25. Seto, M., Kobayashi, Y., Kitao, S., Haruki, R., Mitsui, T., Yoda, Y., Nasu, S., Kikuta, S.: Phys. Rev., B **61**, 11420 (2000)
26. Chumakov, A.I., Sturhahn, W.: Hyperfine Interact. **123/124**, 781 (1999)
27. Alp, E.E., Sturhahn, W., Toellner, T.S.: J. Phys., Condens. Matter **13**, 7645 (2001)
28. Alp, E.E., Sturhahn, W., Toellner, T.S.: Hyperfine Interact. **135**, 295 (2001)
29. Alp, E.E., Sturhahn, W., Toellner, T.S., Zhao, J., Hu, M., Brown, D.E.: Hyperfine Interact. **144/145**, 3 (2002)
30. Handke, B., Kozłowski, A., Parliński, K., Przewoźnik, J., Ślęzak, T., Chumakov, A.I., Niesen, L., Kąkol, Z., Korecki, J.: Phys. Rev., B **71**, 144301 (2005)
31. Brand, R.A.: Nucl. Instrum. Methods Phys. Res., B **28**, 398 (1987)
32. Sturhahn, W.: J. Phys., Condens. Matter **16**, S497 (2004)
33. Sturhahn, W.: Hyperfine Interact. **125**, 149 (2000)
34. Chien, C.L., Unruh, K.M.: Phys. Rev., B **25**, 5790 (1982)
35. Unruh, K.M., Chien, C.L.: Phys. Rev., B **30**, 4968 (1987)
36. Ghafari, M., Brand, R.A., Keune, W.: Hyperfine Interact. **42**, 931 (1988)
37. Ghafari, M., Gonser, U., Wagner, H.G., Naka, M.: Nucl. Instrum. Methods **199**, 197 (1982)
38. Ryan, D.H., Ström-Olsen, J.O., Muir, W.B., Cadogan, J.M., Coey, J.M.D.: Phys. Rev., B **40**, 11208 (1989)
39. Nasu, S., Gonser, U., Preston, R.S.: J. Phys., Colloq. **1**, 385 (1980)
40. Sturhahn, W., Kohn, V.G.: Hyperfine Interact. **123/124**, 68 (1974)
41. Schulze, G.E.R.: Metallphysik, p. 68. Springer, Berlin Heidelberg New York (1974), in German

Hyperfine Interact (2006) 170:47–54
DOI 10.1007/s10751-006-9473-4

Nuclear Resonance Vibrational Spectroscopy (NRVS) of Fe–S model compounds, Fe–S proteins, and nitrogenase

Stephen P. Cramer · Yuming Xiao · Hongxin Wang · Yisong Guo · Matt C. Smith

Published online: 24 January 2007

Abstract We have used nuclear resonance vibrational spectroscopy (NRVS) to examine the nature of the Fe–S unit. Specifically, vibrational characteristics have been determined, and through incremental steps in model system complexity, applied to analysis of the enzyme nitrogenase. This stepwise strategy demonstrates NRVS as a viable bioinorganic tool, and will undoubtedly increase the application of synchrotron spectroscopy to biological problems.

Key words ferredoxin · rubredoxin · vibrational spectroscopy · synchrotron · Mössbauer

1 Introduction

Iron–sulfur species are ubiquitous in nature [1]; among their responsibilities are the enzymatic redox transformations of CO_2, H_2, and N_2 [2]. These substrates, being likely foundations of Earth's early atmosphere, have led Huber and Wächterhäuser to postulate that life may have evolved via the catalytic function of iron–sulfur metalloclusters [3]. The base component of these metalloclusters is the 2Fe2S rhombus, from which many of the larger clusters can be constructed, not just stereochemically, but also literally, as in the case of nitrogenase (N_2ase), the enzyme responsible for the reduction of dinitrogen to ammonia [4]. N_2 fixation is the key step in the nitrogen cycle [5, 6], and this biological ammonia synthesis is responsible for about half of the protein available for human consumption. In *Azotobacter vinelandii* (*Av*) the Mo-dependent N_2ase that accomplishes this reaction uses electrons from an Fe_4S_4 cluster in a 63 kDa Fe protein (Av2) to reduce a 230 kDa $\alpha_2\beta_2$ MoFe protein (Av1). Within Av1, an Fe_8S_7 'P-cluster' supplies electrons to the active site $MoFe_7S_9$ 'FeMo-cofactor', sometimes called the 'M-center.' This is extractable into organic

S. P. Cramer (✉) · Y. Xiao · H. Wang · Y. Guo · M. C. Smith
Department of Applied Science, University of California, Davis, CA 95616, USA
e-mail: spjcramer@mac.com

S. P. Cramer · H. Wang
Physical Bioscience Division, Lawrence Berkeley National Laboratory, Berkeley, CA 94720, USA

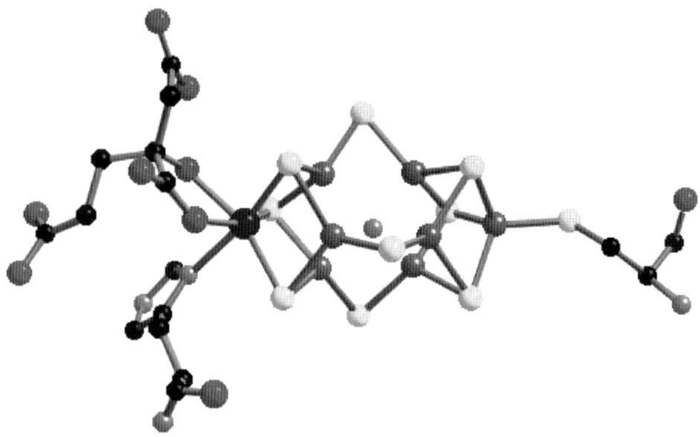

Fig. 1 *Ball and stick* representations of *A. vinelandii* N_2ase FeMo-cofactor, illustrating central location of proposed light atom (PDB code 1M1N) [11]

solvents as 'FeMoco' (Fig. 1) [7]. Protein-bound FeMo-cofactor is ligated by a cysteine on one end, and on the other by histidine and homocitrate ligands [8–10].

We have examined model systems and iron–sulfur proteins (including N_2ase and FeMoco) by the technique of nuclear resonance vibrational spectroscopy (NRVS). The theory behind this technique is described in detail in other articles in this issue, however it is suffice to say here that the above technique is element specific, and through suitable isotopic labeling allows for focus upon the metal atoms within the catalytic machinery. With this spectroscopic tool we hope to address questions that are beyond the reach of protein crystallography; furthermore, important catalytic intermediates are not always the species that can be crystallized, and crystallized species may not necessarily be part of the catalytic cycle. To validate our NRVS data collection and analysis procedures, we initially recorded spectra for a number of model compounds. We took an *aufbau* approach, starting with simple systems and gradually building up to complex clusters. The early work simulated the obtained spectra using Urey–Bradley force fields; the later studies were complemented by DFT calculations. Model compound and small protein studies remain critical for (1) improving experimental methods, (2) refining transferable UBFF force fields, (3) testing DFT predictions, and (4) expanding a database for interpretation of enzyme spectra.

2 Results

Our work in the area of NRVS is summarized below, beginning with the more simple model systems and culminating in the more involved enzymatic constructs. These latter investigations are complicated by (1) the relative diluteness of the iron–sulfur centers within the proteinous environment, (2) the presence of several species, for example [4Fe–4S] clusters that serve as electron conduits, (3) and the greater complexity afforded by a decrease in symmetry, that is often concomitant with the increased number of atoms in the active site. However, even with these factors, evidence is presented that clearly demonstrates that NRVS is entirely applicable as a method for biological studies.

2.1 Model compounds

$[FeH(D)_6]^{4-}$ This was our first NRVS project, conducted largely as a feasibility study, with an eye on hydrogenase projects in the future. We compared these isotopomers to see if

Fig. 2 NRVS of $[Fe(H/D)_6]^{4-}$

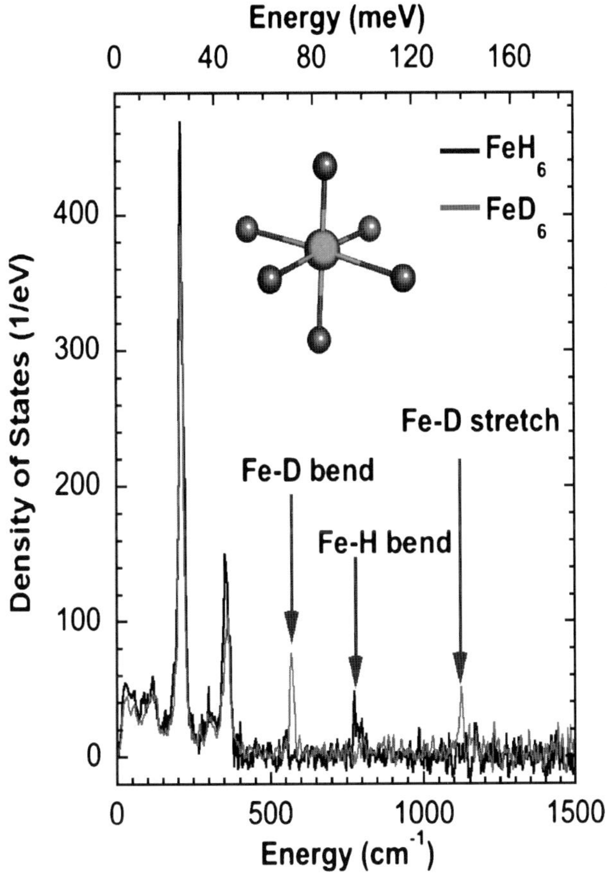

Fe-H/D stretches, which involve little Fe motion, would be visible by NRVS. As shown in Fig. 2, Fe–D modes were visible in 2002 [12]. Of note was the surprising strength of the H–Fe–H and D–Fe–D bend modes.

$R(FeCl_4)$ and $[NEt_4]_2[Fe_2S_2Cl_4]$ These compounds were analyzed by a combination of NRVS, Raman, and IR spectroscopies (data not shown). The D_{2h} symmetry of the $[Fe_2S_2Cl_4]^{2-}$ anion with its centre of inversion makes the IR and Raman modes mutually exclusive, thus we were able to observe and assign 16 of the 18 normal modes for this species. Additionally, we were able to observe a change in $[FeCl_4]^-$ symmetry with counterion, and the acoustic phonon modes for all of these samples [13].

$[Fe_4S_4(SPh)_4]^{2-}$ This cluster was examined as a 'simple' model for 4Fe ferredoxins, and as a test for our ability to observe ^{36}S isotope effects. Owing to the low symmetry and a solid-state phase transition at 233 K, the NRVS spectra turned out to be surprisingly complex. Incorporation of ^{36}S into the bridging S positions produced ~9 cm^{-1} shifts in some NRVS bands (Fig. 3). These shifts were reproducible with DFT calculations [14].

$[Fe_6N(CO)_{15}]^{3-}$ This cluster was examined as one of the few models available for interstitial N in a 6-Fe cage, and as a test for our ability to observe $^{15}N/^{14}N$ isotope effects.

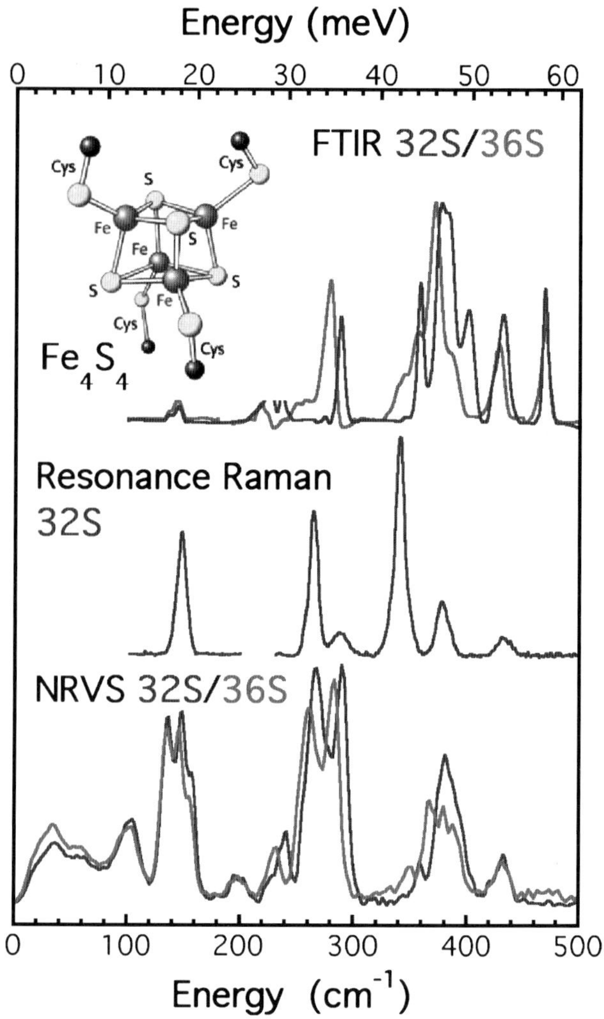

Fig. 3 NRVS (*bottom*) of $[Fe_4S_4(SPh)_4]^{2-}$ with ^{32}S (*blue*) or ^{36}S (*red*) vs. Raman (*middle*) and IR (*top*)

We observed several strong breathing modes for the 6Fe prism, as well as the expected Fe–C stretching bands. A special characteristic of complexes with interstitial 'X' is the presence of 'shake' modes for the X vibrating within the metal cage. The shake modes are very strong in the IR, and give rise to large $^{15}N/^{14}N$ shifts (Fig. 4). We have now seen these shifts in the NRVS as well.

2.2 Protein studies

Rubredoxin We chose *Pyrococcus furiosus* Rd as our first protein because of its 'simple' single FeS_4 center (Fig. 5) [15]. The results turned out to be more complex than expected; there has been a long-standing debate over delocalization of Fe–S modes in Rd. Resonance Raman work had shown an asymmetric Fe–S stretch region divided into three bands near 350–370 cm^{-1} [16]; in our Raman spectra we observed these and additional bands out to 440 cm^{-1}.

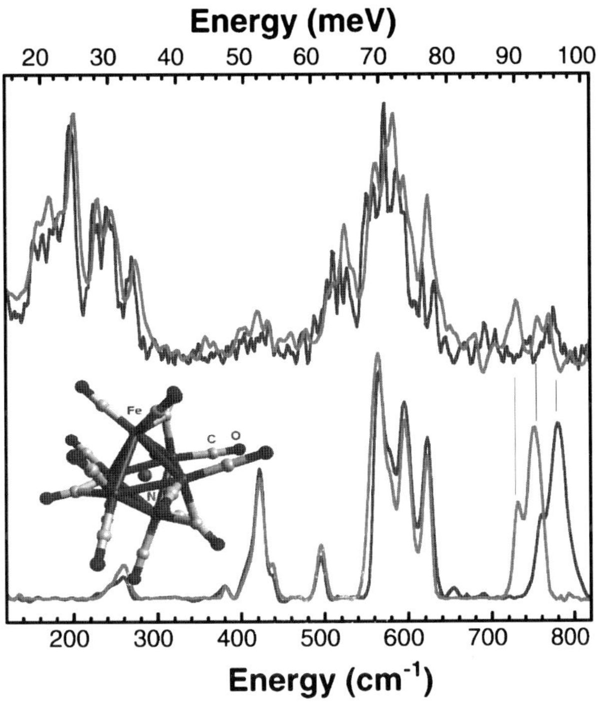

Fig. 4 NRVS (*top*) and IR (*bottom*) for ^{14}N (–) vs. ^{15}N (–) $[Fe_6N(CO)_{15}]^{3-}$

The NRVS was also very broad in this region, suggesting that stretching modes are strongly coupled with protein side chain motion. A model with five-atom chains extending from the Fe site was required to quantitatively reproduce the Fe–S stretch region—quite similar to Goddard's 'chromophore in protein' model [17].

2Fe and 4Fe ferredoxins We recorded NRVS and resonance Raman for *Aquifex aeolicus* (*Aa5*) and *Rhodobacter capsulatus* (*Rc6*) 2Fe Fds. The Fe PVDOS reveals a strong and broad low frequency region (Fig. 5), suggesting that the Fd bend modes are highly delocalized and mixed with larger scale peptide motions. Raman spectra for ^{57}Fe/^{36}S-substituted *Rc6* exhibit isotope shifts that provide additional constraints on the normal mode analyses. We followed with measurements on a *Pf* 4Fe Fd using the D14C mutant with all Cys side chains [18]. Compared to 2Fe Fds, the 4Fe cluster shows less coupling with the protein matrix.

Nitrogenase and FeMoco Together with Bill Newton and Karl Fisher, we examined samples of ^{57}Fe-enriched Av1, a Δ*nifE* Av1 mutant containing only P-cluster, and isolated FeMoco (Fig. 6). The difference spectrum between Av1 and Δ*nifE* Av1 represents the protein-bound FeMo-cofactor. In this data and in the FeMoco NRVS, the catalytic site exhibited a strong signal near 190 cm^{-1}, where conventional Fe–S clusters have weak NRVS. This intensity was ascribed to cluster breathing modes whose frequency is raised by an interstitial atom. A variety of Fe–S stretching modes are also observed between 250 and 400 cm^{-1}. The spectra were reasonably well simulated both by empirical UBFF force fields and by DFT calculations [19].

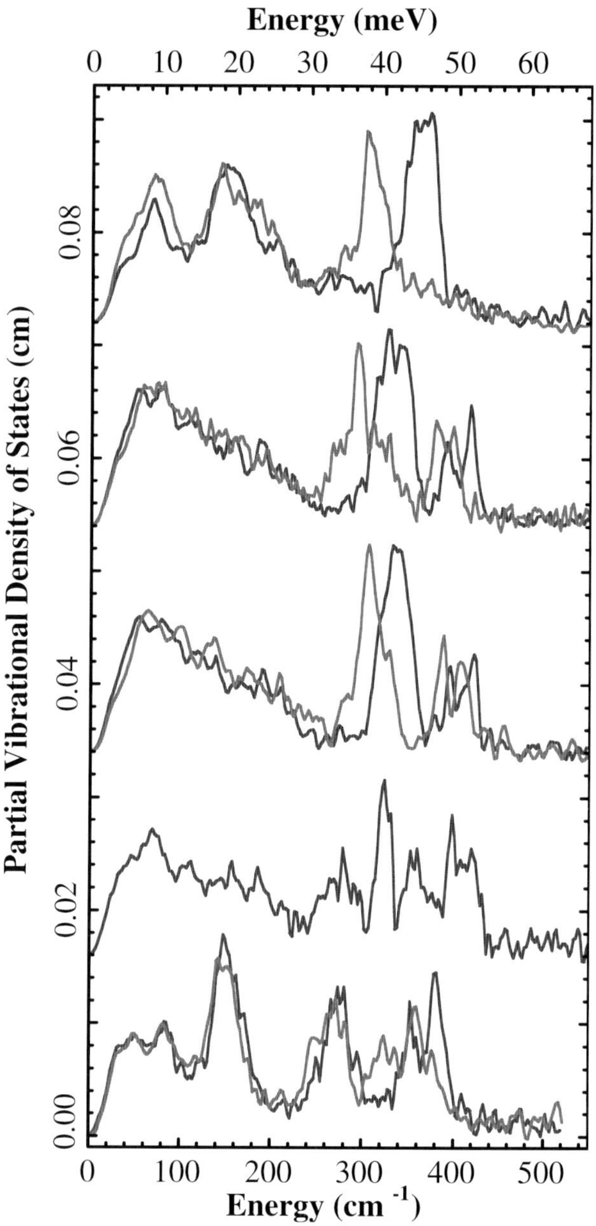

Fig. 5 NRVS of Fe–S proteins. *Top to bottom*, oxidized (red) vs. reduced (blue): (a) *Pf* Rd, (b) 2Fe *Aa5* Fd, (c) 2Fe *Rc6* Fd, (d) 2Fe Rieske protein, (e) 4Fe D14C *Pf* Fd

It is worth emphasizing that this is the first vibrational information ever obtained about the FeMo-cofactor or P-cluster metals. There is a wealth of information to be obtained from these spectra, especially when combined with DFT calculations and isotopic labeling. We are obviously quite excited about the potential of these experiments.

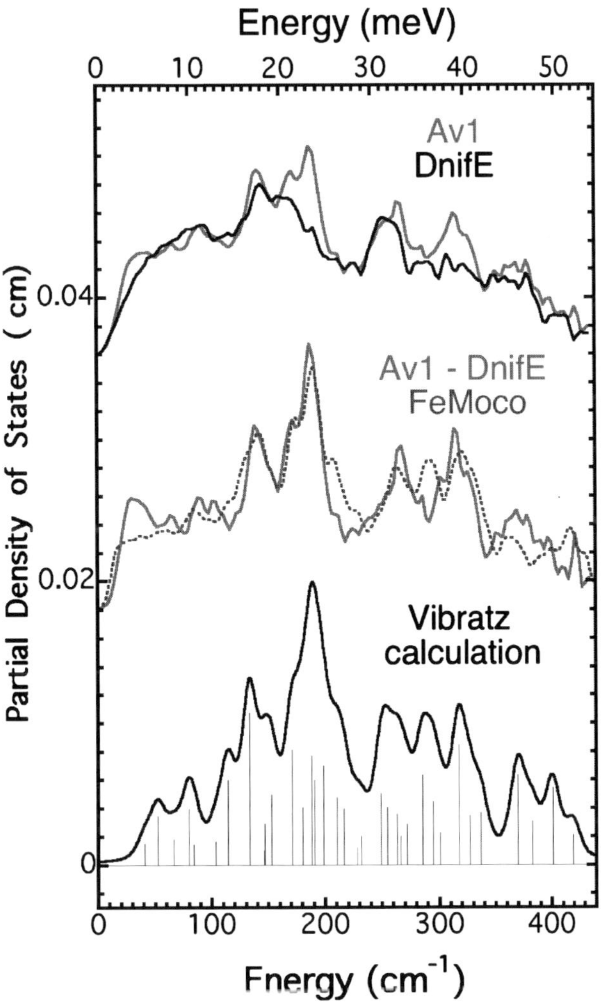

Fig. 6 NRVS of MoFe N_2ase vs. the P-cluster only ΔNifE, MoFe *minus* ΔNifE vs. FeMoco. Vibratz simulation of FeMoco

2.3 Summary

The results presented above unequivocally illustrate that synchrotron spectroscopies have a role to play in the ever increasingly complex attack on unraveling the secrets of metalloenzymes, and no doubt with continued future development will become more routine and readily available.

3 Experimental

NRVS spectra were recorded at beamline 3-ID at the APS, Illinois, and at beamline 09-XU at SPring-8, Japan as previously described [13, 15]. PVDOS were calculated using PHOENIX [20].

Acknowledgements We thank E. Ercan Alp, Jiyong Zhao and Wolfgang Sturhahn at the APS and Yoshitaka Yoda at SPring-8 for assistance with our experiments, and Mike Adams, Jacques Meyer, Thomas Rauchfuss, William Newton, Karl Fisher, and Roberto Della Pergola for samples. This work was funded by NIH grants GM-44380 (SPC), GM-65440 (SPC), and the DOE, Office of Biological and Environmental Research (SPC). The Advanced Photon Source is supported by the DOE, Office of Basic Energy Sciences. SPring-8 is supported by JASRI—the Japan Synchrotron Radiation Research Institute.

References

1. Rees, D.C.: Ann. Rev. Biochem. **71**, 221–246 (2002)
2. Rees, D.C., Howard, J.B.: Science. **300**, 929–931 (2003)
3. Huber, C., Wächterhäuser, G.: Science. **281**, 670–672 (1998)
4. Agar, J.N., Zheng, L.M., Cash, V.L., Dean, D.R., Johnson, M.K.: J. Am. Chem. Soc. **122**, 2136–2137 (2000)
5. Igarashi, R.Y., Seefeldt, L.C.: Crit. Rev. Biochem. Mol. Biol. **38**, 351–384 (2003)
6. Dos Santos, P.C., Igarashi, R.Y., Lee, H.I., Hoffman, B.M., Seefeldt, L.C., Dean, D.R.: Accts. Chem. Res. **38**, 208–214 (2005)
7. Pickett, C.J., Vincent, K.A., Ibrahim, S.K., Gormal, C.A., Smith, B.E., Best, S.P.: Chem. Eur. J. **9**, 76–87 (2003)
8. Kim, J.K., Rees, D.C.: Nature. **360**, 553–560 (1992)
9. Bolin, J.T., Campobasso, N., Muchmore, S.W., Morgan, T.V., Mortenson, L.E.: The structure and environment of the metal clusters in the nitrogenase MoFe protein from *Clostridium pasteurianum*. In: Stiefel, E.I., Coucouvanis, D., Newton, W.E. (eds.) Molybdenum Enzymes, Cofactors, and Model Systems, pp. 186–195. American Chemical Society, Washington, D.C. (1993)
10. Mayer, S.M., Lawson, D.M., Gormal, C.A., Roe, S.M., Smith, B.E.: J. Mol. Biol. **292**, 871–891 (1999)
11. Einsle, O., Tezcan, F.A., Andrade, S.L.A., Schmid, B., Yoshida, M., Howard, J.B., Rees, D.C.: Science. **297**, 1696–1700 (2002)
12. Bergmann, U., Sturhahn, W., Linn, D.E., Jenney, F.E., Adams, M.W.W., Rupnik, K., Hales, B.J., Alp, E.E., Mayse, A., Cramer, S.P.: J. Am. Chem. Soc. **125**, 4016–4017 (2003)
13. Smith, M.C., Xiao, Y.M., Wang, H.X., George, S.J., Coucouvanis, D., Koutmos, M., Sturhahn, W., Alp, E.E., Zhao, J.Y., Cramer, S.P.: Inorg. Chem. **44**, 5562–5570 (2005)
14. Xiao, Y., Koutmos, M., Case, D.A., Coucouvanis, D., Wang, H., Cramer, S.P.: Dalton Trans. **18**, 2192–2201 (2006)
15. Xiao, Y.M., Wang, H.X., George, S.J., Smith, M.C., Adams, M.W.W., Jenney, F.E., Sturhahn, W., Alp, E.E., Zhao, J.O., Yoda, Y., et al.: J. Am. Chem. Soc. **127**, 14596–14606 (2005)
16. Czernuszewicz, R.S., Kilpatrick, L.K., Koch, S.A., Spiro, T.G.: J. Am. Chem. Soc. **116**, 7134–7141 (1994)
17. Qiu, D., Dasgupta, S., Kozlowski, P.M., Goddard, W.A., Spiro, T.G.: J. Am. Chem. Soc. **120**, 12791–12797 (1998)
18. Brereton, P.S., Duderstadt, R.E., Staples, C.R., Jonhson, M.K., Adams, M.W.W.: Biochem. **38**, 10594–10605 (1999)
19. Xiao, Y., Fischer, K., Smith, M.C., Newton, W., Case, D.A., George, S.J., Wang, H., Sturhahn, W., Alp, E.E., Zhao, J., Yoda, Y., Cramer, S.P.: J. Am. Chem. Soc. **128**, 7608–7612 (2006)
20. Sturhahn, W.: Hyperfine Interact. **125**, 149–172 (2000)

Hyperfine Interact (2006) 170:55–60
DOI 10.1007/s10751-006-9474-3

On Spin Hamiltonian fits to Mössbauer spectra of high-spin Fe(II) porphyrinate systems

Charles E. Schulz · Chuanjiang Hu · W. Robert Scheidt

Published online: 21 December 2006

Abstract Fits to Mössbauer spectra of high-spin iron(II) porphyrinates have been applied to the Fe(II) model compounds octaethylporphyrin(1,2-dimethylimidazole) and tetra-paramethoxyporphyrin(1,2-dimethylimidazole). Mössbauer spectra have been measured on these compounds at 4.2 K in large applied fields. Spin Hamiltonians were used for fitting both the electronic and nuclear interactions. The fits are done by adjusting the Hamiltonian parameters to simultaneously minimize the total χ^2 for three different applied fields. In order to get best fits, the EFG tensor need to be rotated relative to the ZFS tensor. A comparative sensitivity analysis of their Spin Hamiltonian parameters has also been done on the ZFS parameters D, and the EFG asymmetry parameter η. The best fits suggest that both systems definitely have a negative quadrupole splitting, and that largest EFG component is tilted far from the z-axis of the ZFS tensor, which is likely to be near the heme normal.

Key words porphyrin · high-spin · electronic field gradient (EFG) · Mössbauer spectroscopy

The understanding of the detailed electronic state of Fe(II) in porphyrin compounds is of considerable interest, as they are models for the heme proteins myoglobin and hemoglobin in their active oxy- and deoxy-states. Mössbauer spectroscopy has been important in a number of studies of deoxyhemoglobin and model compounds that have similar iron coordination geometries [1–4].

C. E. Schulz (✉)
Department of Physics, Knox College, Galesburg, IL 61401, USA
e-mail: cschulz@knox.edu

C. Hu (✉) · W. R. Scheidt
Department of Chemistry and Biochemistry, University of Notre Dame, Notre Dame, IN 46556, USA
e-mail: scheidt.1@nd.edu

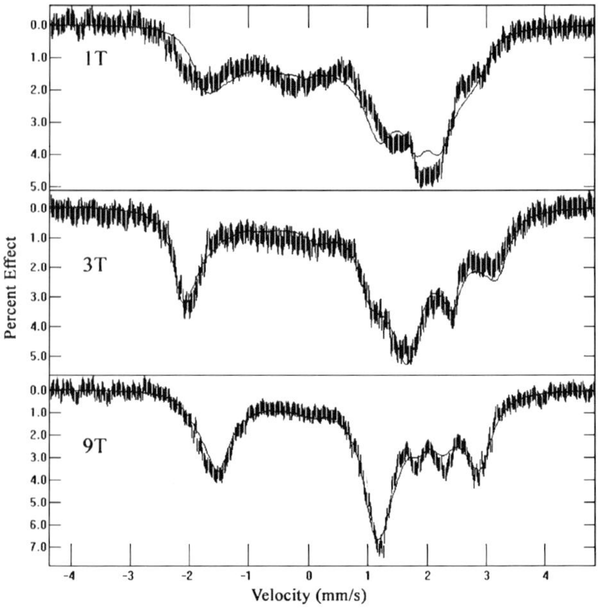

Figure 1 Mössbauer spectra of **1** at 4.2 K taken in applied fields as shown. The field direction is parallel to the gamma ray beam.

As they are non-Kramers (integer spin) systems, such proteins and model compounds are considerably more challenging to characterize than Fe(III) systems, for which EPR can routinely provide corroborating information. For high-spin ferrous iron in coordination compounds, the ground $S = 2$ multiplet is split by spin-orbit coupling mixing the ground multiplet with excited states [5]. The size of this splitting (zero-field splitting, or ZFS) is a measure of the nearness of excited states. The ZFS is commonly described by a Spin Hamiltonian, the simplest form of which is

$$H = D\left[S_z^2 - 1/3\, S(S+1)\right] + E\left(S_x^2 - S_y^2\right) + \mu_B \cdot \vec{H} \cdot \tilde{g} \cdot \vec{S} \quad (1)$$

The parameters D and E are the axial and rhombic zero-field spitting parameters, can be calculated in second-order perturbation theory, and are of order λ^2/Δ, where λ is the spin-orbit coupling parameter and Δ the energy of an excited crystal field state being mixed into the ground orbital. The last term represents the Zeeman interaction of the iron spin with an external magnetic field H. For the characterization of the Mössbauer spectra of a system, the nuclear Hamiltonian must be considered as well. The iron nucleus will experience an electric quadrupole interaction with the local electric field gradient (EFG), a magnetic hyperfine interaction with the spin S, and nuclear Zeeman interaction as well. Thus the nuclear Hamiltonian will look like

$$H_n = \vec{I} \cdot \tilde{Q} \cdot \vec{I} + \vec{S} \cdot \tilde{A} \cdot \vec{I} - g_n \beta_n \vec{H} \cdot \vec{I} \quad (2)$$

where Q is proportional to the EFG and A is the magnetic hyperfine tensor.

There are clearly many parameters available to fit a Mössbauer spectrum. These include the D and E parameters, two independent components of the traceless EFG tensor, and the three components of the A tensor. In principle, the three components of the g tensor are available fit parameters as well, though they can be constrained by perturbation equations that depend on the D and E parameters and λ [6].

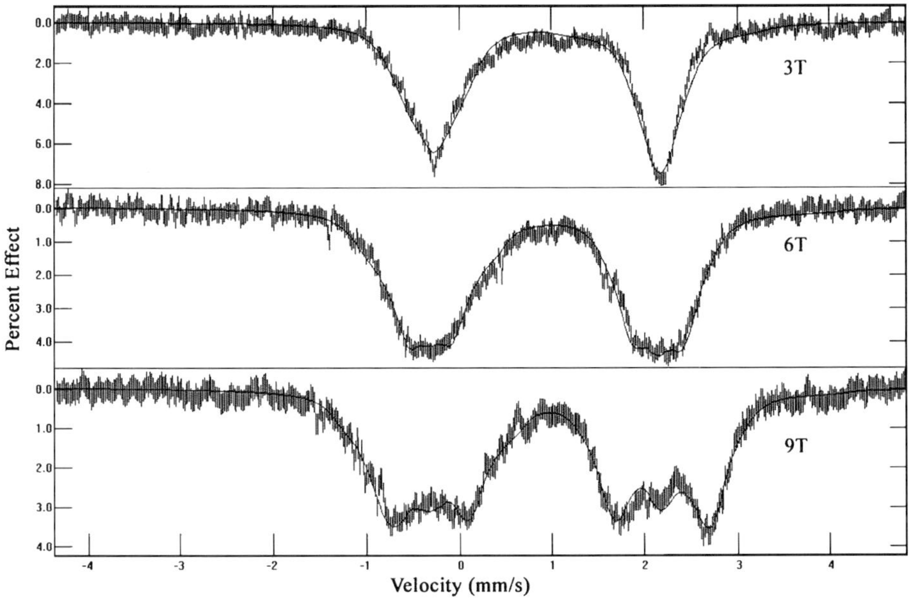

Figure 2 Mössbauer spectra of **2** at 4.2 K taken in applied fields as shown. The field direction is parallel to the gamma ray beam.

But a complete description of the ZFS may not be quite as simple as that. There exist possible axial and rhombic contributions to the ZFS energy that are proportional to electronic spin operator to the fourth power. An argument can be made that these fourth-order contributions should be small relative to the second-order terms D and E, but one should be careful with such facile arguments. There is also no a priori reason to assume that the principal axes of the g, A, and Q tensors are all collinear. The relative orientations of two tensors' principal axes can be described by a set of three Euler angles. Thus there are six such angle parameters that may be important for a correct physical characterization of a system's Mössbauer spectra. Our experience has shown that many high-spin ferrous systems require these rotation parameters for successful fits.

Given so many possible parameters, it is crucial to have multiple constraints to maximize the possibility that in adjusting parameters to fit experimental Mössbauer data, the final fit parameters have some connection with physical reality. In this work we will expand on this important issue by reporting a study of the uniqueness of Mössbauer spectra fits as applied to the Fe(II) model compounds octaethylporphyrin(1,2-dimethylimidazole) (hereafter **1**) and tetraparamethoxyporphyrin(1,2-dimethylimidazole) (hereafter **2**).

Applied-field Mössbauer spectra of Fe(II) systems with seemingly similar iron coordination environments can be strikingly different. Figures 1 and 2 illustrate this point. System **1** shows a broad magnetic splitting at 4.2 K in a large applied field, while **2** has only weakly resolved structure under identical conditions. This makes the point that high-spin Fe(II) is a highly sensitive probe of its electrostatic, magnetic, and coordination environment. The solid lines in Figures 1 and 2 are our best fits based on the Spin Hamiltonians, Eqs. 1 and 2. The fits are done by adjusting the

Table I Constrained tensor rotation summary

System	A-tensor	EFG tensor	Best χ^2
1	Free to rotate	Free to rotate	0.906
1	Fixed parallel to ZFS	Free to rotate	0.950
1	Fixed parallel to ZFS	Fixed parallel to ZFS	1.128
2	Free to rotate	Free to rotate	0.582
2	Fixed parallel to ZFS	Free to rotate	0.648
2	Fixed parallel to ZFS	Fixed parallel to ZFS	0.760

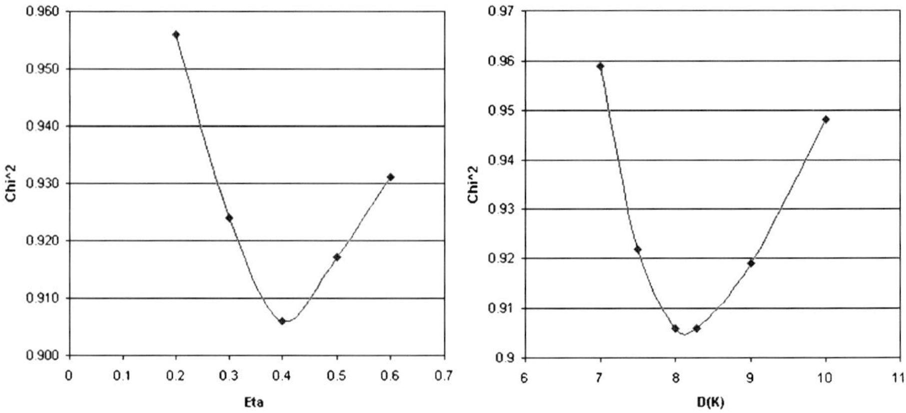

Figure 3 Increase in χ^2, resulting from (*left*) constraining η to a particular value while optimizing all other parameters while fitting **1**, and (*right*) a similar graph for the ZFS axial field parameter D. The solid lines are a guide for the eye.

Hamiltonian parameters to simultaneously minimize the total χ^2 for three different applied fields. The minimization was done use the Downhill Simplex algorithm [7].

Best fits for **1** and **2** could not be achieved without allowing the EFG tensor to be rotated relative to the ZFS tensor. Naturally, the fits improved further upon allowing rotation of the magnetic hyperfine tensor A. The reduced χ^2 values shown in Table I demonstrate this. While the degradation of χ^2 upon constraining the rotations is not extremely large, it should be noted that the "non-rotated-tensor" fits to **1** also resulted in non-standard EFG tensors ($|\eta| > 1$), in itself equivalent to one or more 90° rotations of the EFG.

In the Mössbauer literature for which values of Spin Hamiltonian parameters have been published, it is often the case that no uncertainties are given for those parameters. The uncertainty of the fit values will depend on the volume in parameter space (surrounding the minimum that has been found) over which χ^2 remains close to the minimum. Parameters are likely to be more tightly constrained by the fits for sets of spectra that have sharply-defined features (as in **1**) than those which do not (as in **2**). Thus this pair of compounds provides us the opportunity for a comparative sensitivity analysis of their Spin Hamiltonian parameters. Among the parameters most of interest for defining the characteristics of the iron site are the ZFS parameters D and E, and the EFG asymmetry parameter η, equal to $(V_{xx} - V_{yy})/V_{zz}$. Figures 3 and 4 show the results of our sensitivity analysis. As expected, χ^2 increases

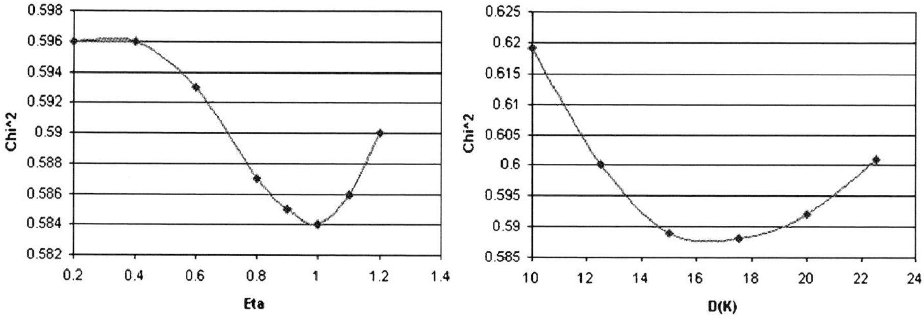

Figure 4 The same analysis as Figure 3, but on **2**.

Table II Spin Hamiltonian parameters

Complex	**1**	**2**
$D(\text{cm}^{-1})$	5.8	12.2
$E(\text{cm}^{-1})$	0.92	3.8
ΔE_Q, (mm/s)	−2.19	−2.44
η	0.42	1.0
δ_{Fe}, mm/s	0.92	0.95
$A_{xx}/g_N\beta_N$(T)	−2.68	−14.0
$A_{yy}/g_N\beta_N$(T)	−12.7	−1.55
$A_{zz}/g_N\beta_N$(T)	−8.1	−44.5
Euler angles Q→D (°)	104, −64, 154	1, 101, 91
Euler angles A→D (°)	−77, −73, 86	0, 5, 0

much more quickly as either D or η is moved away from the optimal values for system **1**. For system **2**, the quality of the simultaneous fits to the three spectra appear visually indistinguishable for values of D ranging from 10 to over 20 K (7 to 14 cm^{-1}).

So while we can conclude that D is much larger for **2** than for **1**, the precise value in **2** remains quite uncertain. Similarly, η is almost completely undetermined. Tightening the range of acceptable values of D for **2** would require additional and complementary information, such as Mössbauer spectra taken in large fields at higher temperatures (for which the fast relaxation limit applies), from magnetic susceptibility measurements, or under fortuitous circumstances from integer-spin EPR measurements.

The values for the Spin Hamiltonian parameters that minimized χ^2 for these systems are given in Table II. We note briefly what these fits tell us about the two samples. Both systems definitely have a negative quadrupole splitting, which means that both of them have a negative value for largest component of the EFG. In both systems, that largest EFG component is tilted far from the z-axis of the ZFS tensor, which is likely to be near the heme normal. Although the value of D is indefinite for **2**, it likely to be much larger than for **1**. This means that it has lower-lying excited orbitals, resulting from a weaker axial ligand field, which spin-orbit interactions then mix more strongly into the ground orbital/spin quintet. Finally, we note that both E/D and the η parameters are larger for **2**, so by both measures its iron is in a much more rhombic environment than is **1**.

A further step in understanding these compounds would be to attempt to do a detailed ligand-field calculation, which would check that the values for the A tensor parameters are plausible and consistent with the ZFS parameters yielded by the fits. In addition, it would be useful to compare these two systems with studies of other Fe(II) porphyrin models having constrained imidazole ligands, to look for systematic correlation of the Mössbauer results with some characteristic of the molecular structure. Such studies are well underway.

References

1. Maeda, Y., Harami, T., Trautwein, A., Gonser, U.: Z. Naturforscher **31B**, 487 (1976)
2. Kent, T.A., Spartalian, K., Lang, G., Yonetani, T.: Biochim. Biophys. Acta **490**, 331 (1977)
3. Kent, T.A., Spartalian, K., Lang, G.: J. Chem. Phys. **71**, 4899 (1979)
4. Hu, C., Roth, A., Ellison, M.K., An, J., Ellis, C.M., Schulz, C.E., Scheidt, W.R.: J. Am. Chem. Soc. **127**, 5675 (2005)
5. Abragam, A., Bleaney, B.: Electron Paramagnetic Resonance of Transition Ions. Oxford University Press, London, UK (1970)
6. Zimmerman, R.: Ph.D. Thesis, University of Erlangen-Nürnberg, Germany (1973)
7. Press, W.H., Teukolsky, S.A., Vetterling, W.T., Flannery, B.P.: Numerical Recipes in Fortran. Cambridge University Press, Cambridge, UK (1986)

Hyperfine Interact (2006) 170:61–66
DOI 10.1007/s10751-006-9471-6

Mössbauer effect of ^{151}Eu in europium salen complex nanoparticles

C. I. Wynter · D. E. Brown · M. Iwunze · S. G. Sobel ·
Leopold May · F. W. Oliver · A. Adeweymo

Published online: 9 January 2007

Abstract The use of Mössbauer spectroscopy to investigate nanoparticle systems is fairly widespread. In this study, nanoparticles of a europium salen complex, NH_4 Eu $(salen)_2$, were prepared by the sol–gel method, and its properties investigated using infrared and Mössbauer spectroscopy. The size of the nanoparticles was measured using an atomic force microscope. The Debye temperature of the nanoparticles was determined from the Mössbauer spectrum and compared with the Debye temperature of the salen complex previously reported. The infrared spectral data support the Mössbauer spectroscopic results.

Key words europium salen complex nanoparticle · Mössbauer spectroscopy · infrared spectra · AFM

1 Introduction

The synthesis, lattice dynamics, and magnetic properties of $NH_4Eu\ (salen)_2$, have recently been reported [1]. Also the lattice dynamics of europium benzoate, fluoride, and oxalate

C. I. Wynter (✉)
Nassau Community College, Garden City, NY 11530-6793, USA
e-mail: wynterc@ncc.edu

D. E. Brown
Physics Department, North Illinois University, DeKalb, IL 60439, USA
e-mail: DEBrown@physics.niu.edu

M. Iwunze · A. Adeweymo
Chemistry Department, Morgan State University, Baltimore, MD 21251, USA

S. G. Sobel
Chemistry Department, Hofstra University, Hempstead, NY 11549, USA

L. May
Department of Chemistry, The Catholic University of America, Washington, DC 20064, USA

F. W. Oliver
Physics Department, Morgan State University, Baltimore, MD 21251, USA

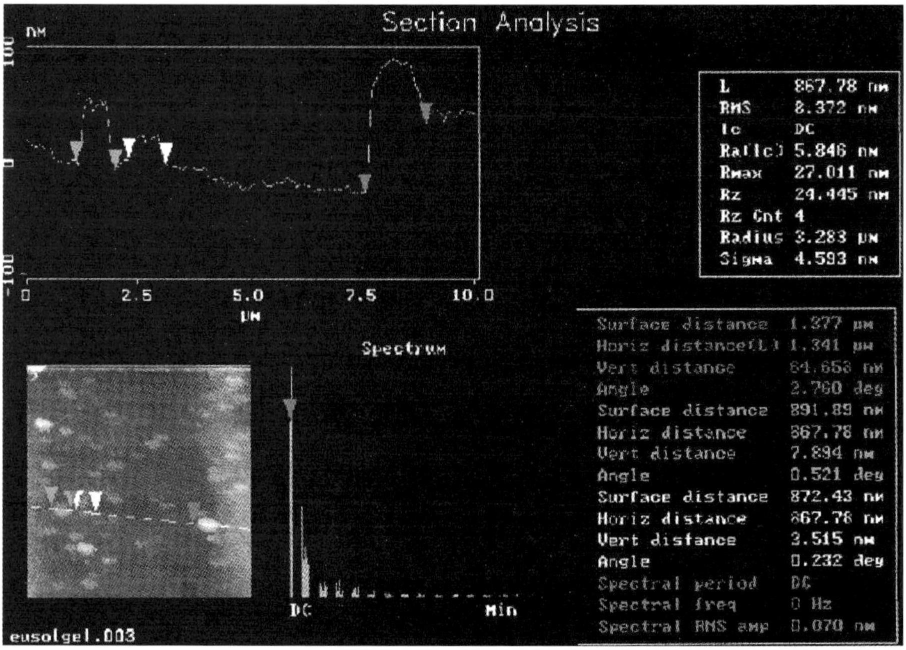

Fig. 1 AFM picture of a Eu-salen nanoparticle

were studied [2]. We now report on the preparation of nanoparticles of this compound by the sol–gel method [3] as well as the Debye temperature (θ_D) of these nanoparticles as compared to the salen complex. The structure of the nanoparticles was examined using the atomic force microscopy, and the infrared spectra of the nanoparticles are compared to the spectra of the salen complex and related to the Mössbauer spectral data.

2 Experimental

The NH_4Eu (salen)$_2$ complex was prepared as previously described [1]. Nanoparticles of $NH_4Eu(salen)_2$ were prepared inside a sol–gel matrix. The sol was prepared by mixing tetramethylorthosilicate (TMOS), formamide, and water in a volume ratio of 3:3:5. A quantity of 1.0 ml of 0.10 M HNO_3 was added as a catalyst to aid the condensation and polymerization processes. The resulting sol was typically cloudy and was stirred using a magnetic stirrer until a clear isotropic solution was obtained. This sol was stored in a refrigerator at a temperature of 4±0.2°C until it was ready for use.

The encapsulation of complex was made with a solution of 0.0505 g of the complex dissolved in 5.0 ml of 0.10 M HNO_3. A quantity of 1.0 ml of this solution was mixed with 3.0 ml of the previously prepared sol solution. The sol containing the complex was allowed to gel and age in air for 2 days after which it was dried in an oven at 68.0°C overnight.

For preparation for AFM measurements, the cylindrical sol–gel in which the complex was encapsulated was mounted on the AFM instrument and scans taken at different angles of 0.232, 0.406, 0.521, and 2.76° were obtained. The average horizontal distance at small

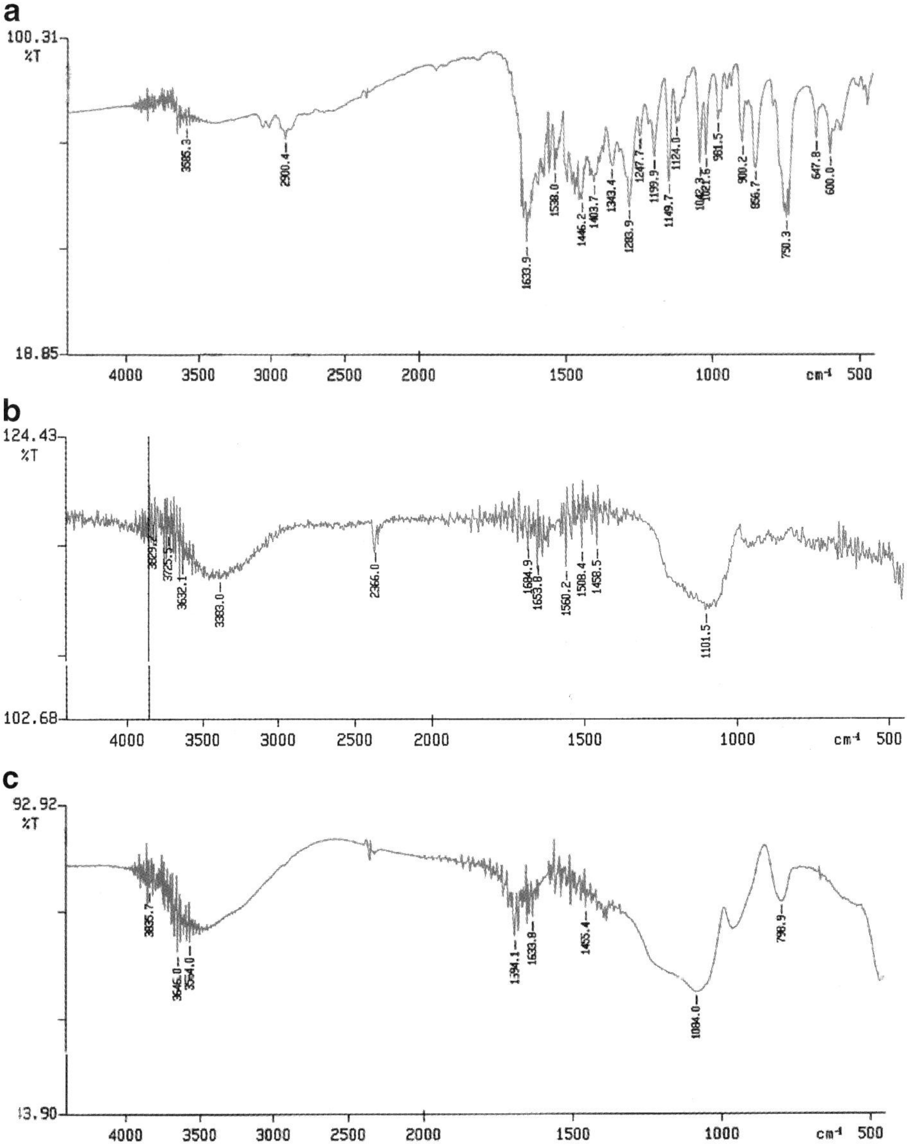

Fig. 2 **a** Infrared spectrum of $NH_4Eu(salen)_2$. **b** Infrared spectrum of the sol–gel. **c** Infrared spectrum of the nanoparticles

angles (0.232, 0.406, and 0.521°) was taken as the particle size. A Digital Instrument Atomic Force Microscope (model DI 300) was used.

For the infrared spectral measurements, the nanoparticles were dried in an oven at 70.0°C for 24 h after which it was pulverized in a mortar and thoroughly ground to a fine powder with KBr. A pellet of this mixture in which the ratio of Eu-salen to KBr was 1:10 by mass was made. All infrared spectra were obtained using a FT-IR spectrophotometer (Perkin–Elmer model Spectrum RX FT-IR) with a KBr pellet in the reference beam.

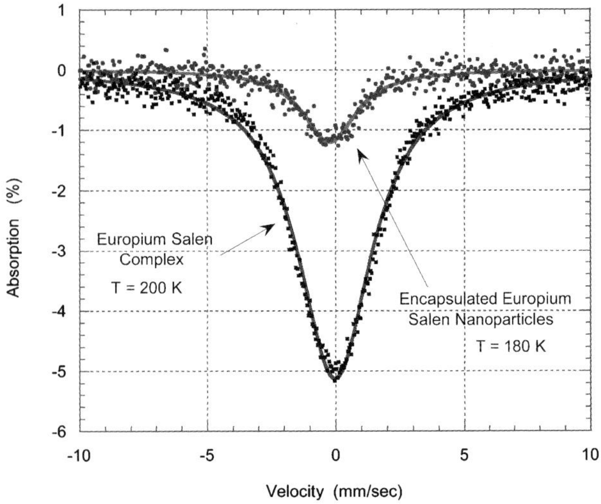

Fig. 3 Mössbauer transmission data for the encapsulated europium salen nanoparticles and the europium salen complex

The ^{151}Eu Mössbauer spectra of the nanoparticles were obtained using a 100 mCi $^{151}Sm_2O_3$ source, and the absorber had a concentration of roughly 180 mg/cm^2 of encapsulated europium salen nanoparticles. A Canberra Germanium detector and a Janis SVT-400 flowing vapor cryostat were used to obtain the data. The velocity was calibrated with an iron foil at room temperature, and a EuF_3 sample having a concentration of 5.5 mg/cm^2 was also measured at room temperature as a standard. The Mössbauer spectra of each of the two compounds were measured at temperatures ranging from 5 to 180 K in a constant acceleration mode.

3 Results and discussion

Figure 1 shows the AFM picture of a Eu-salen nanoparticle taken at the angle of 0.232°. The AFM picture of the nanoparticle taken at 0.406 and 0.521° are not shown, but its surface and horizontal distances are not different from those obtained at the angles of 0.232 and 0.521°. An average of the surface distance of the Eu-salen nanoparticle taken at angles of 0.232, 0.460 and 0.521° is 868 nm. The aspect ratio of surface to the horizontal distance is 1:1.

The significance of this is that the synthesized nanoparticle is spherically shaped with an approximate radius of 434.0 nm. The surface distance of the wide-angle measurement of 2.76° shown in Fig. 1, is 1.377 μm and its horizontal distance is 1.341 μm. This confirms that the aspect ratio of this particle is approximately 1:1.

The infrared spectra of the nanoparticles and its components are shown in Fig. 2. Figure 2a, b, and c are the spectrum of the complex, the sol–gel, and the nanoparticles, respectively. There are three peaks indicative of salen presence, 1634 cm^{-1} ($v - C = N$), 1400 cm^{-1} (NH_4^+) and 799 cm^{-1} (Ph breathing) in the nanoparticles. The phenyl breathing mode is increased in frequency from 750 cm^{-1}, in the spectrum of the complex alone, possibly due to confinement in the matrix, which may make breathing modes more 'stiff' in the nanoparticles. A new band appears at about 980 cm^{-1}. The sol–gel broad band, which was 1050–1350 cm^{-1}, with a peak at 1084 cm^{-1} and an unclear shoulder around 1260 cm^{-1},

Table 1 Mössbauer spectral results for the nanoparticles

Temperature, K	δ^{a}	FWHM[b]	A^{c}	Ln A
5	0.26(1)	2.43(3)	1.00(2)	0
15	0.25(1)	2.48(3)	0.99(1)	−0.0098
25	0.28(1)	2.48(5)	0.92(2)	−0.0804
35	0.26(2)	2.50(6)	0.86(2)	−0.1518
45	0.28(2)	2.59(6)	0.81(2)	−0.2131
55	0.24(2)	2.53(6)	0.71(2)	−0.3402
65	0.25(2)	2.53(6)	0.69(2)	−0.3767
75	0.26(2)	2.44(5)	0.59(1)	−0.5299
90	0.28(1)	2.50(5)	0.55(1)	−0.6041
120	0.29(2)	2.43(7)	0.41(1)	−0.8946
150	0.26(2)	2.44(8)	0.30(1)	−1.1932
180	0.30(3)	2.5(1)	0.25(1)	−1.4024

[a] Isomer shift relative to EuF_3, mm/s (to get the isomer shifts relative to iron foil, subtract 0.65 from these values)

[b] Mössbauer linewidth (full width at half maximum), mm/s

[c] Normalized area under spectrum

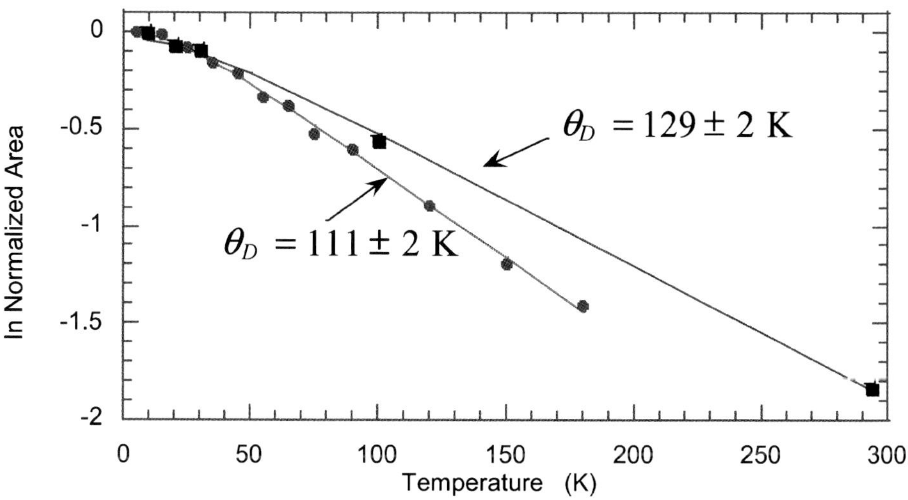

Fig. 4 Temperature dependence of the natural log of the normalized area under the Mössbauer transmission curves for the europium salen nanoparticles (*circles*) and the europium salen complex (*squares*). The area was normalized to unity at the lowest measured temperature

is modified to a wide peak at 1080 cm^{-1} with a more defined shoulder at 1350 cm^{-1}. These changes could be due to overlay of low-frequency salen bands with the sol–gel absorption, or it could be indicative of interaction of the sol–gel with the complex, or the absorption of the sol–gel obscures the absorption due to the complex.

Figure 3 shows typical Mössbauer data of the nanoparticles and the salen complex. The absorption of the salen nanoparticles is noticeably weaker than the complex due to both the smaller quantity of europium and the lower Debye temperature. Table 1 is a summary of the ^{151}Eu Mössbauer data of the nanoparticles at varying temperatures. The line widths

(FWHM) vary little with temperature and are comparable to those previously reported for the macroscopic complex [1] considering that the present results used $^{151}Sm_2O_3$ as a source whereas the previously reported data of the salen complex was with a $^{151}SmF_3$ source. The isomer shifts reported are relative to a measured EuF_3 sample. The small positive isomer shift (approximately 0.3 mm/s) for both the nanoparticles and the complex represents trivalent europium.

The Lamb–Mössbauer factor, or recoilless *f*-factor, is related to the area under the Mössbauer spectrum, *A*, and follows the Debye model for harmonically vibrating oscillators [4]:

$$\ln A \sim \ln f = -\frac{6E_R}{k_B\theta_D}\left(\frac{1}{4} + \left(\frac{T}{\theta_D}\right)^2 \int_0^{\frac{\theta_D}{T}} \frac{x}{e^x - 1}dx\right)$$

where $E_R = \frac{E_\gamma^2}{2M}$ is the recoil energy of the nucleus of the europium atom having mass M, E_γ is the energy of the emitted gamma ray, k_B is the Boltzmann factor, T is the temperature, and θ_D is the Debye temperature. The log *f*-factor is also proportional to the mean square displacement of the vibrating nucleus, Δx^2 : $\log f \sim -k^2\Delta x^2$, where k is the wave vector of the emitted gamma ray. Thus, as the area increases, the, *f*-factor increases (to a theoretical limit of 1), and the vibrational amplitude decreases (the material becomes a more rigid oscillator).

Plots of the logarithm of the area (ln *A*) versus temperature (T) for the complex and the nanoparticles are given in Fig. 4. A fit of the area under the curves using the entire temperature range for both particle sizes gives a Debye temperature of $\theta_D = 111 \pm 2$ K for the nanoparticles and a value of $\theta_D = 129 \pm 2$ K for the complex. This latter θ_D agrees within experimental error with our previously stated value of 133±5 K for the bulk sample where the high temperature Debye approximation, $T \geq \frac{1}{2}\theta_D$, was used [1].

The θ_D of 111 K for the salen nanoparticles is significantly smaller than that of 129 K for the salen complex. This suggests that the europium bonds are slightly less rigid in the nanoparticles than in the complex. This might be associated with the possible bond between the complex and gel as shown from the infrared spectra. In reality, the europium salen complex forms a more rigid, fixed lattice of vibrating europium atoms than the lattice of europium atoms for the encapsulated nanoparticles. This may be indicative of europium salen rattling in their polymer cages for the nanoparticles and thus have larger vibrational amplitudes. A systematic study of Debye temperature with nanoparticle size will be investigated in the future to provide further information on the behavior of europium atoms in nanoparticles.

Acknowledgements Work at Northern Illinois University was supported by the Department of Education.

References

1. Wynter, C.I., Ryan, D.H., May, L., Oliver, F.W., Brown, E., Hoffman, E., Bernstein, D.: Industrial applications of the Mössbauer effect. In: Garcia M., Marco J.F., Plazaola F. (eds.) American Institute of Physics: New York, p. 300, (2005)
2. Wynter, C.I., Ryan, D.H., Taneja, S.P., May, L., Oliver, F.W., Brown, D.E., Iwunzie, M.: Hyperfine Int., DOI 10.1007/s10751-006-9315-4
3. Yan, S.F., Geng, J.X., Chen, J.F., Yin, L., Zhou, Y.C., Zhou, E.L.: J. Cryst. Growth **262**, 415 (2004)
4. Greenwood, N.N., Gibb, T.C.: Mossbauer Spectroscopy. Chapman and Hall, London, (1971)

Hyperfine Interact (2006) 170:67–74
DOI 10.1007/s10751-006-9472-5

Velocity scales for Mars Mössbauer data

David G. Agresti · M. Darby Dyar · Martha W. Schaefer

Published online: 11 January 2007

Abstract Raw Mössbauer data acquired from surface rocks and soils by two Mars Exploration Rovers (MERs) are being made available on-line to the general science community for independent analysis and interpretation. The data are released as counts per channel, so a velocity scale must be defined. The Windows-based computer program *MERView* does this in two steps. In the first, a scale proportional to the true velocity is derived by adjusting the magnitudes of the drive error signal and a phase shift. This report focuses on the second step, in which mm/s units are assigned to the proportional scale by comparing laboratory-measured α-iron peak positions with those of spectra of the internal MER calibration targets. We contrast *MERView*-derived velocity scales for different MER measurement conditions with the velocity scales provided on-line by the MER team, and conclude that independent velocity calibration is generally desirable.

Key words Mössbauer data · Mars · velocity calibration · methodology · *MERView*

1 Introduction

Two Mössbauer spectrometers (MIMOS II) [1] have been operating on the surface of Mars since January 2004 as part of the instrument suite of the Mars Exploration Rovers (MERs) Spirit and Opportunity (http://www.athena.cornell.edu/). They have been acquiring spectra of surface rocks and soils in order to contribute to a better understanding of Mars' geochemical

D. G. Agresti (✉)
Department of Physics, University of Alabama at Birmingham, Birmingham, AL 35294-1170, USA
e-mail: agresti@uab.edu

M. D. Dyar
Department of Astronomy, Mount Holyoke College, 50 College St., South Hadley, MA 01075, USA
e-mail: mdyar@mtholyoke.edu

M. W. Schaefer
Department of Geology and Geophysics, E235 Howe–Russell, Louisiana State University, Baton Rouge, LA 70803, USA
e-mail: schaefer@geol.lsu.edu

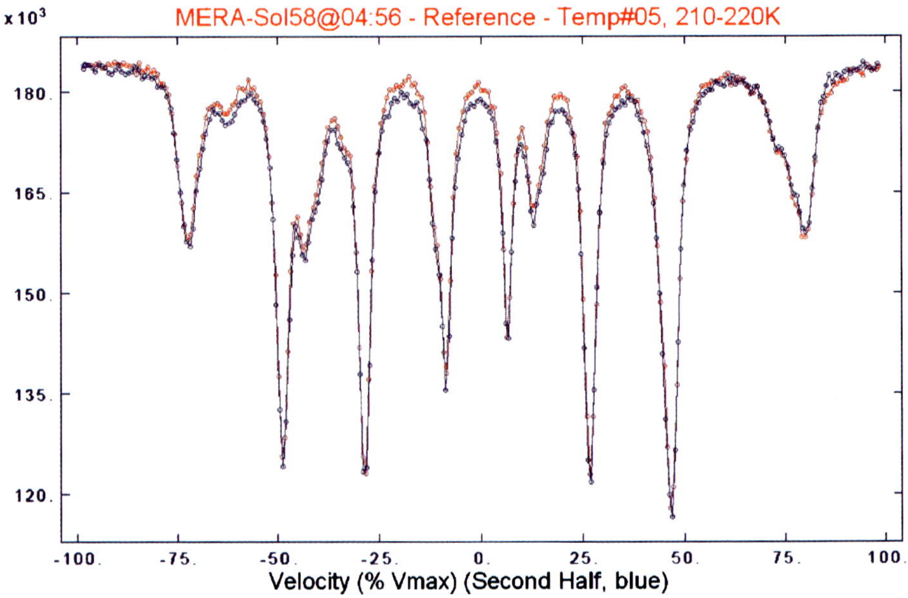

Fig. 1 Reference spectrum overlapped at mid channel and plotted versus velocity. Second half is below the first half

history, in particular the nature of aqueous processes that may have played a role in defining the observed surface mineralogy. Analyses of spectral data by the MER Mössbauer team, with inferred mineralogy and interpretations in the geologic context, are presented in several reports [e.g. 2–4]. Given the importance of a full assessment of the data by outside investigators, the general science community is provided with the opportunity for independent analysis by regular releases of the MER data, at roughly 90-day intervals, to two principal web sites, the MER Analyst's Notebook at http://anserver1.eprsl.wustl.edu/ and the Planetary Data System (PDS) Geosciences Node at http://pds-geosciences.wustl.edu/.

Mössbauer spectra for each rover are recorded in five detectors in up to thirteen 10-K wide temperature windows (<180 K to >290 K), resulting in up to 65 distinct Mössbauer spectra for each sol (Mars day) on which data are recorded. Four detectors record the sample spectra, namely backscatter spectra of surface material, rock, soil, or dust, while detecting 6.4-keV Fe Kα X-rays and 14.4-keV ^{57}Fe γ photons. The fifth detector records in transmission mode the reference spectrum of an internal calibration target consisting of an enriched α-Fe foil overlaid with enriched α-Fe_2O_3 and Fe_3O_4. The raw Mössbauer data for a particular sol, essentially as received from Mars, are embedded in a single 160-kB binary-coded file, the Experimental Data Record (EDR), which also includes other pertinent information such as detector energy spectra, temperature data, and a drive error signal. The structure of the EDR file and the nature of the binary coding are described in a Software Interface Specification document, available on line. Most important for this report is that the Mössbauer data are provided as counts versus channel, requiring calibration of the velocity scale to mm/s versus channel, which may be accomplished with use of the drive error signal and the properties of the calibration target.

Shortly after the first MER data release in August 2004, we introduced a new Windows-based computer program, *MERView*, to provide convenient access to the raw Mössbauer data and a capability to calibrate the velocity scale. In its initial version, *MERView* extracted

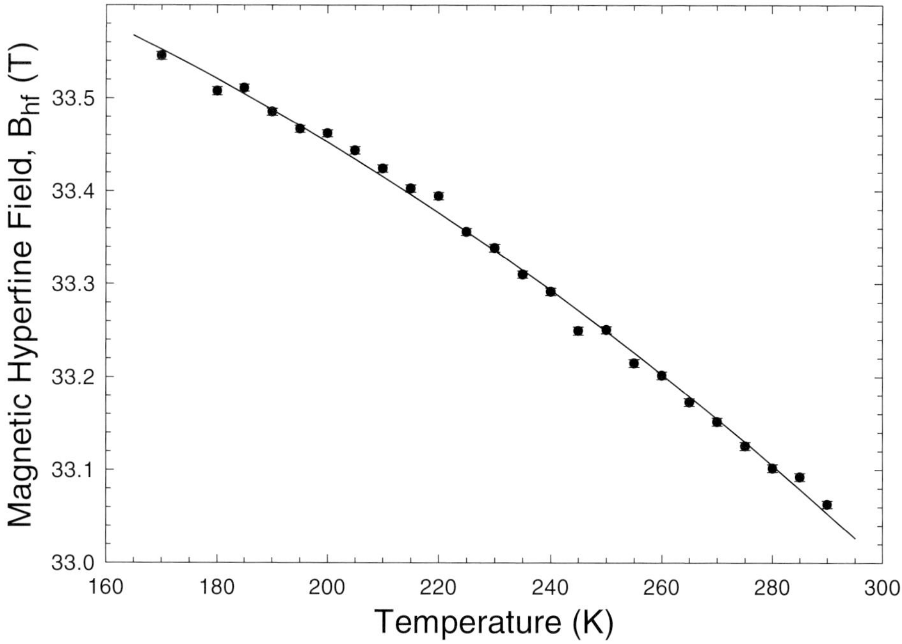

Fig. 2 Measured values of B_{hf} for an α-Fe foil as function of temperature. The *solid line* is a parabolic fit to the data

the individual spectra from the EDR file, converted the data from binary to decimal, provided tabular displays and output files, and produced full-screen graphical displays for rapid survey of the entire EDR data set [5]. Subsequently, *MERView* was enhanced to correct for the rather substantial non-linearity in the velocity by including a channel offset (phase shift) and adjusting the amount of the EDR-provided drive error signal added to a linear, V-shaped waveform, while optimizing overlap of the two halves of the reference spectrum [6]. The resulting scale is proportional to the true velocity scale, but not yet in mm/s. This report presents the final step in the calibration process, namely that of assigning mm/s units to the proportional velocity scale.

2 Velocity calibration

Figure 1 shows a typical 512-channel reference spectrum for MER-A (Spirit), *corrected for non-linearity and phase shift* (see [6] for methods), and folded about the mid-channel point. The maximum of the underlying V-shaped, linear scale has been arbitrarily set to *Vmax*= 100%. Note that the first half (red, channels 1–256) of the full spectrum is displayed from right to left, while the second half (blue) goes from left to right, in accordance with a roughly V-shaped velocity waveform. As a consequence of drive non-linearity, complementary channels for the two halves are generally not at the same velocity.

The α-Fe component of the reference spectrum, that is, the six deepest peaks of Fig. 1, may be used to calibrate the velocity in mm/s. The hematite and magnetite subspectra are not appropriate, since they vary widely as function of preparation conditions, information about which is lacking for the MER calibration targets. Analysis shows that four of the

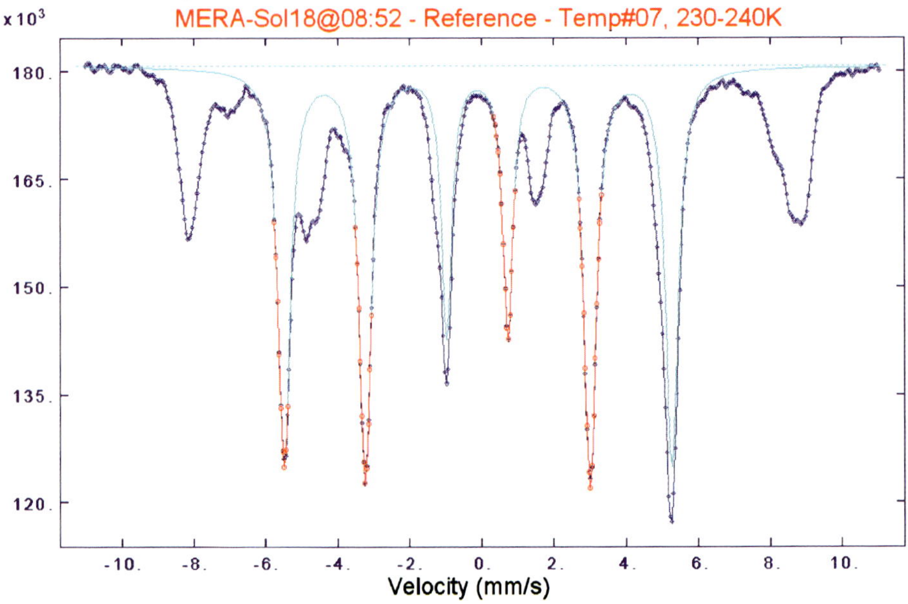

Fig. 3 MER-A reference (*blue*) and same-temperature α-Fe spectrum (*green*) with a computed baseline (*dotted*) and selected parts of four α-Fe peaks (*red*)

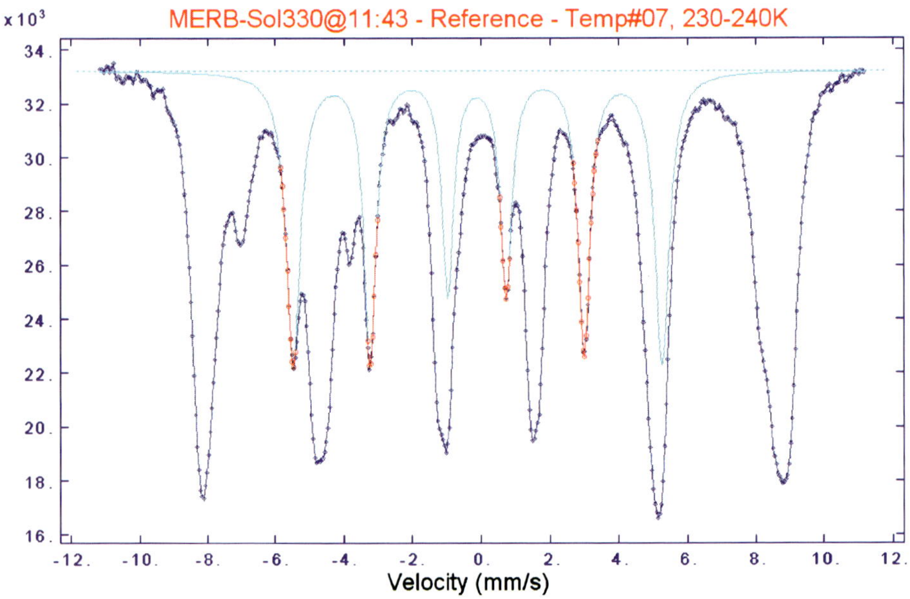

Fig. 4 MER-B reference (*blue*) and same-temperature α-Fe spectrum (*green*) with a computed baseline (*dotted*) and selected parts of four α-Fe peaks (*red*)

α-Fe peaks (1, 2, 4, and 5, counting left to right in the figure) are relatively free of overlap. To establish a laboratory basis for the calibration, we acquired spectra of a 25-μm α-Fe foil (ESPI Metals) from 170–290 K and fit for the ground-state splitting ΔE_g in mm/s. The results, converted as described in [7] to magnetic hyperfine field B_{hf}, are presented in Fig. 2,

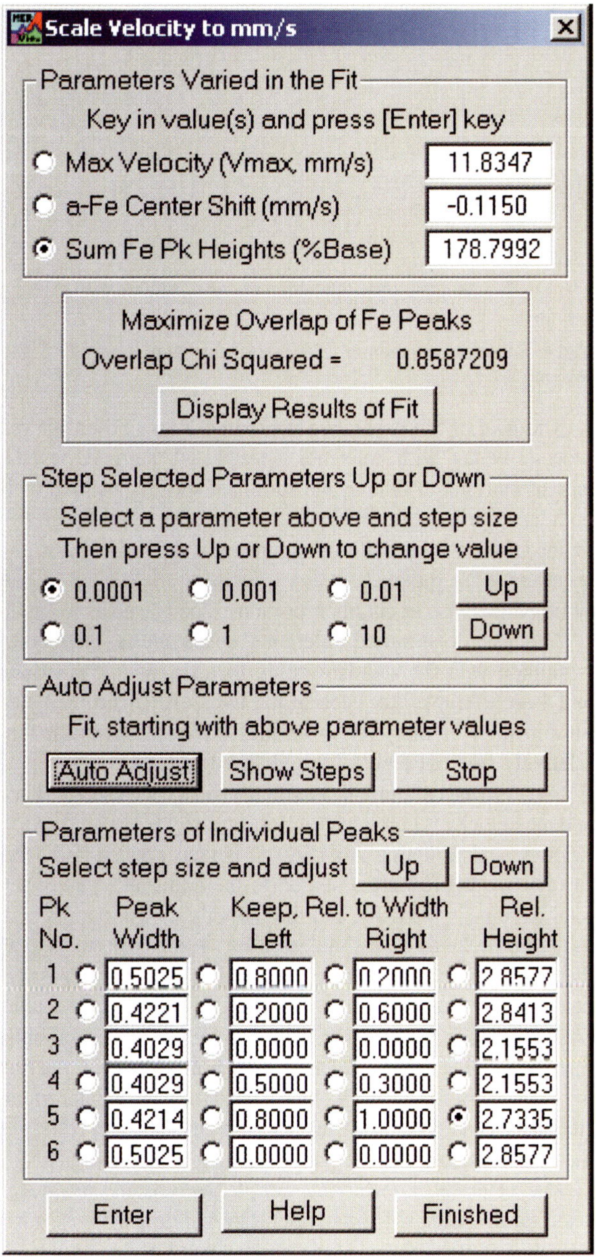

Fig. 5 Velocity scaling dialog box, as it appears after auto adjustment for Fig. 3

along with a parabolic fit to the data. Values for ΔE_g at the mid-point of each MER temperature window, inferred from this statistically-smoothed variation, represent the calibration standard we have used. Since information is lacking about the nature of the α-Fe foils used on the MERs, the absolute precision in *MERView*-derived velocities will be limited to ~0.5% (c.f. [8]; Tom Kent, personal communication).

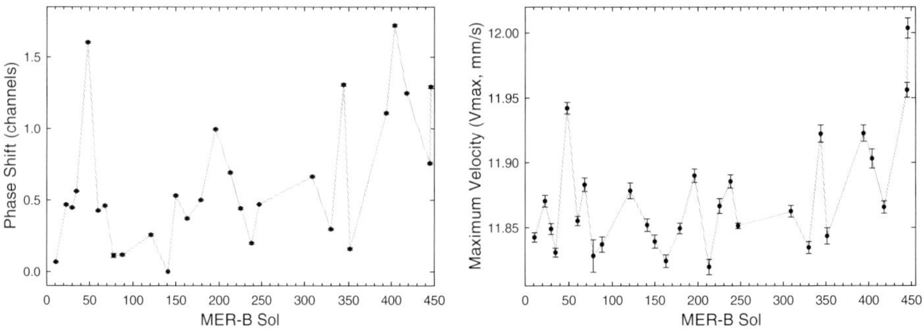

Fig. 6 Sol-to-sol variation of *CShift* (*left*) and *Vmax* (*right*) for MER-B, temperature window 07 (selected sols only)

The goal of the calibration procedure is to find an optimal value for *Vmax* in mm/s. The process works with reference spectra as "summed pseudo data," similar to the typical laboratory result of folding and summing spectra acquired with a linear drive in order to include both halves in a single "folded" spectrum. For a non-linear drive, pseudo velocities must first be defined, here with a uniform increment of 0.4% *Vmax*, or ~10 values per peak width; the raw spectrum is interpolated at each velocity to yield a set of pseudo data [6]; and these are then summed pairwise at common pseudo-velocity values to yield the summed pseudo data.

The pseudo-summed reference spectrum is then compared with an α-Fe spectrum computed with the standard value for ΔE_g at mid window (see above). Shown in Figs. 3 and 4 are examples for each of the two MER instruments at the conclusion of least-squares optimization of the overlap of the reference spectrum with selected portions of the four relatively free α-Fe peaks (highlighted in red).

To begin the calibration, a particular reference spectrum is selected (MER, sol, temperature), and the velocity is corrected for non-linearity [6], resulting in a display like that of Fig. 1. The velocity scaling dialog box (Fig. 5) is then opened and certain pre-selected values (different for the two MERs) are assigned to the adjustable parameters, *Vmax*, α-Fe spectrum centre, total α-Fe peak intensity, and the individual peak widths and relative intensities. The parts of the α-Fe peaks to be fit, indicated by "Keep, Rel. to Width," Left/Right, are also preset, so that the influence of hematite and magnetite subspectra is reduced. Parameters for the unused peaks 3 and 6 are set equal to those of peaks 4 and 1, respectively. If the initial overlap is reasonable, the AutoAdjust button is pressed, the variables are automatically adjusted to minimize "Overlap Chi Squared" (progress shown on the box), and the desired velocity scale is obtained. Before starting AutoAdjust, any parameter value may be entered individually or stepped up and down by a selected increment, either to improve the initial overlap or to study the effect of changing the parameter.

The velocity scale for the sample spectrum is the negative of the associated reference scale for the same sol and temperature window because surface material is interrogated with a separate source placed at the opposite end of the single MIMOS II drive. After calibration, both scales are included on graphical displays and in exported data files.

3 Assessment of velocity scales

For both MER-A and MER-B, the MER team has provided a "Velocity scale for surface measurements," obtainable by selecting "Data Set Documents" from the Analyzer's Notebook web site. As of this writing, these "universal" scales make no allowance for sol-to-sol variation

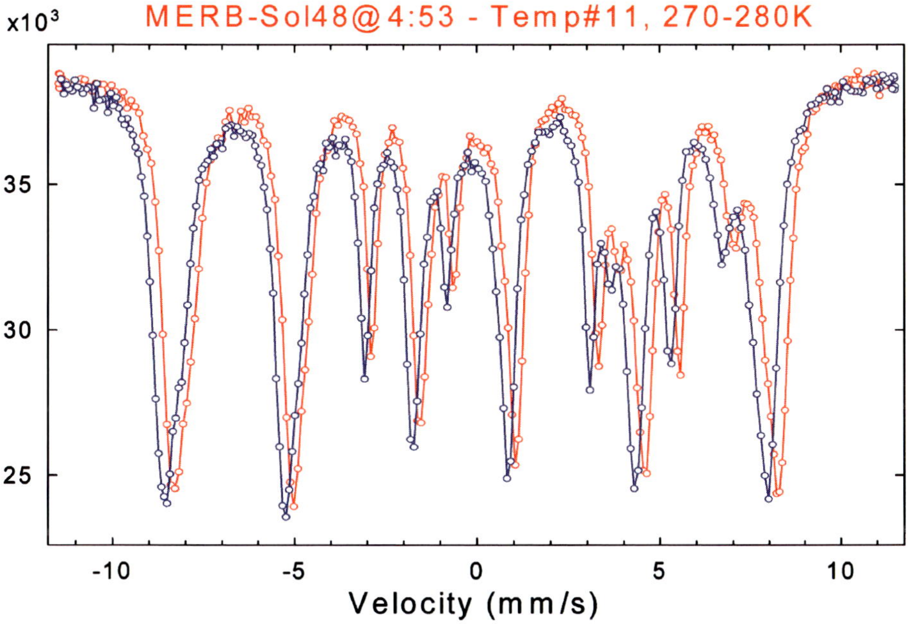

Fig. 7 Raw data for MER-B, sol 48, temperature window 11, plotted versus corresponding team-provided velocity scale

and apply only to temperature windows 3–11. In the following, we will demonstrate that there is sufficient sol-to-sol variation to question the universal applicability of these scales. There are also cases where the scales do not apply at all.

There are three adjustable quantities that directly affect the velocity scale, namely the size of the error signal, *ErrAmp*, the channel shift, *CShift*, and *Vmax*. *MERView* calibration shows that all of these vary from sol to sol, as well as with the temperature. Examples of such variation for *CShift* and *Vmax* are shown in Fig. 6.

In Fig. 7 we have graphed data for one sol where *CShift* is particularly large. As we see, common spectral features from the two halves do not line up at the same velocity, demonstrating the failure of the team-provided scale. Many other examples like this have been found during *MERView* analyses of the MER reference spectra, most not so dramatic as this one. However, to be certain one has a correct scale, velocity calibration should be carried out independently for each spectrum of interest.

Several dozen reference spectra among the two MERs were acquired during temperature windows 2, 12, or 13, which are not included in the team-provided velocities scales. There are also a number of measurements that were made at reduced velocity (*Vmax* ~4 or ~6 mm/s) for better resolution near zero velocity. By definition, the universal scales cannot apply in either of these cases. Finally, many reference spectra exhibit large distortions in the baseline, especially for the lower temperatures, which complicates the non-linearity correction. The reason for this is unknown. These cases should be checked individually to assure proper velocity calibration.

4 Conclusion

Independent velocity calibration of Mössbauer data acquired on the surface of Mars can now be carried out with *MERView*, a new Windows-based computer program. With this capability, we have assessed the general applicability of the team-provided "universal" scales and conclude that, for highest precision and for cases where the scales do not apply, the user should derive a velocity scale for each distinct reference spectrum, that is, for each MER, sol, and temperature window. We have now processed the MER sample spectra currently available, associated *MERView*-derived velocity scales with them, and posted the results at author MDD's web site, http://www.mtholyoke.edu/courses/mdyar/database/. A copy of the *MER-View* program may be obtained by request from author DGA at http://www.agresti@uab.edu.

Acknowledgements We are grateful for support from NASA grants NAG5-12687 and NNG06G130G (to MDD) and NNX06AD93G (to DGA).

References

1. Klingelhöfer, G., Morris, R.V., Bernhardt, B., Rodionov, D., de Souza, P.A. Jr, Squyres, S.W., Foh, J., Kankeleit, E., et al.: J. Geophys. Res. **108** (E12), 8067 (2003). DOI http://dx.doi.org/10.1029/2003JE002138
2. Morris, R.V., Klingelhöfer, G., Bernhardt, B., Schröder, C., et al.: Science **305**, 833–836 (2004)
3. Klingelhöfer, G., Morris, R.V., Bernhardt, B., Schröder, C., et al.: Science **306**, 1740–1745 (2004)
4. Morris, R.V., Klingelhöfer, G., Schröder, C., Rodionov, D.S., et al.: J. Geophys. Res. **111** (E02S13) (2006). DOI http://dx.doi.org/10.1029/2005JE002584
5. Agresti, D.G., Dyar, M.D., Schaefer, M.W.: Lunar Planet. Sci. **36**(1941) (2005) (CD-ROM)
6. Agresti, D.G., Dyar, M.D., Schaefer, M.W.: Hyp. Interact. (2006). DOI http://dx.doi.org/10.1007/s10751-006-9370-x
7. Dyar, M.D., Agresti, D.G., Schaefer, M.W., Grant, C.A., Sklute, E.C.: Ann. Rev. Earth Planet. Sci. **34**, 83–125 (2006). DOI http://dx.doi.org/10.1146/annurev.earth.34.031405.125049
8. Preston, R.S., Hanna, S.S., Heberle, J.: Phys. Rev. **128**, 2207–2218 (1962)

Hyperfine Interact (2006) 170:75–82
DOI 10.1007/s10751-006-9469-0

Nanowires of iron oxides embedded in SiO_2 templates

N. A. Fellenz · J. F. Bengoa · S. G. Marchetti · F. R. Sives ·
S. J. Stewart · R. C. Mercader

Published online: 21 December 2006

Abstract To attain the complete filling of the channels of MCM-41 with magnetite and maghemite, we have tried out an alternative method to the incipient wetness impregnation. The mesoporous material was instilled with a Fe-carrying organic salt after subjecting the matrix to a silylation treatment. Thus, a solid of 7.7 wt.% iron-loaded MCM-41 was obtained. Different subsequent thermal treatments were used to produce γ-Fe_2O_3 or Fe_3O_4. The Mössbauer and magnetic results show that after this method, the as-prepared composite displays a size-distribution of magnetic particles. It is mainly made up of fine particles that display a superparamagnetic relaxation at room temperature and get blocked at ≈42 K for the AC susceptibility time-scale measurements both for γ-Fe_2O_3 and Fe_3O_4 particles. For both samples, about 24% of larger iron-containing phases are magnetically blocked at room temperature. For the Fe_3O_4 particles, this fraction undergoes the Verwey transition at about 110 K; in addition, there is a minor Fe (III) fraction that remains paramagnetic down to 4.2 K.

Key words γ-Fe_2O_3 · Fe_3O_4 · MCM-41-supported magnetic nanoarrays · nanoparticles magnetism · silylated MCM-41

1 Introduction

Highly ordered mesoporous materials can be used conveniently as templates to confine magnetic compounds in low-dimensional arrays [1, 2]. The properties that exhibit the arrays of nanowires and nanotubes so obtained are currently intensively investigated because of the non-traditional phenomena observed and because of the variety of their possible applications [3]. In the last decade a matrix-mediated approach has been used as a means to

N. A. Fellenz · J. F. Bengoa · S. G. Marchetti
Departamento de Química, Facultad de Ciencias Exactas, Universidad Nacional de La Plata, CONICET, CINDECA, CICPBA, 47 N° 257, 1900 La Plata, Argentina

F. R. Sives · S. J. Stewart · R. C. Mercader (✉)
Departamento de Física, Facultad de Ciencias Exactas, Universidad Nacional de La Plata, IFLP-CONICET, CC. 67, 1900 La Plata, Argentina
e-mail: mercader@fisica.unlp.edu.ar

obtain magnetic nanoclusters of special properties and active supported catalysts [4–8]. The most common nanofabrication techniques, like laser ablation [9], lithography [10] or supercritical fluid phase inclusion [11], are relatively complex and expensive. Some time ago [12], we set about using the simpler dry impregnation of SiO_2 templates toward obtaining nanotubes and nanowires of iron oxides.

Although the regular hexagonal parallel channels of MCM-41 (average diameters less than ≈5 nm) offers itself as an almost ideal support, the irregular distribution of silanol groups on the channel walls hinders the uniform filling of the pores. We have shown that hematite (α-Fe_2O_3) does not fill homogeneously the MCM-41 pores after impregnation by incipient wetness with an aqueous solution of iron nitrate [13, 14].

Considering this result, we have attempted a different approach towards filling completely the matrix pores. Prior to the impregnation, we carried out a treatment to homogenize the distribution of silanol groups on the inner walls of the matrix. Because the channels are hydrophobic except at the areas where the silanol groups are sited, a previous silylation treatment can turn the inner surface into hydrophobic, allowing the pores filling with an organometallic iron salt in an organic solvent. In this work, we show preliminary results on the impregnation of the ≈3 nm MCM-41 channels with a Fe-carrying organic salt after subjecting the support to a silylation treatment; afterwards, different thermal treatments applied to the solid yielded magnetite or maghemite particles inside its pores.

2 Experimental

The MCM-41 was prepared according to the methodology proposed by Ryoo and Kim [15], using sodium silicate in alkaline medium and cetyl-trimethylammonium chloride as surfactant. Both for the maghemite-and magnetite-loaded samples, the matrix, MCM-41 of specific surface area of 940 m^2/g and average pore diameter of 2.8 nm, was dehydrated at 573 K and $p_{vacuum}<10^{-3}$ Torr by heating for 3 h, and was silylated afterwards. The silylation treatment was carried out with hexamethyl-disilazane (HMDS) in toluene (0.05 M) inside a glove box in N_2 atmosphere. This solution was added to the dehydrated MCM-41 (20 cm^3/g solid). The mixture was heated at 393 K for 90 min with stirring. Then it was filtrated and washed with toluene to remove the HMDS excess. Finally, the solid was dried in an oven at 333 K for 16 h and impregnated with the $Fe(AcAc)_3$ solution in benzene with a concentration enough to produce an iron loading of about 5 wt.%.

After impregnation, the solid was heated at 13 K/min from room temperature to 683 K and annealed at this temperature in a closed vessel for 20 h. In previous assessments these had proved to be the best preparation conditions to decompose bulk $Fe(AcAc)_3$ into Fe_3O_4.

A separate fraction was subjected to a subsequent thermal treatment carried out heating from room temperature up to 580 K at 2.4 K/min in air flow (150 cm^3/min) and keeping it for 2 h a the highest temperature. This treatment completely converted the Fe_3O_4 into γ-Fe_2O_3 (maghemite).

The resulting composites were characterized by X-ray diffraction (Cu $K\alpha$ radiation) and atomic absorption spectroscopy. The Mössbauer spectra at 298 and 22 K were taken in transmission geometry with a constant acceleration spectrometer. A source of ^{57}Co in a Rh matrix of nominally 50 mCi was used. The isomer shifts, δ, were calibrated against a metallic α-Fe foil at room temperature. The temperature between 22 and 298 K was varied using a Displex DE-202 closed cycle cryogenic system. The 4.2 K Mössbauer spectrum of the Fe_3O_4-loaded sample was taken using a sinusoidal velocity waveform, in a cryostat where both the source and the sample were held at liquid helium temperature. No liquid

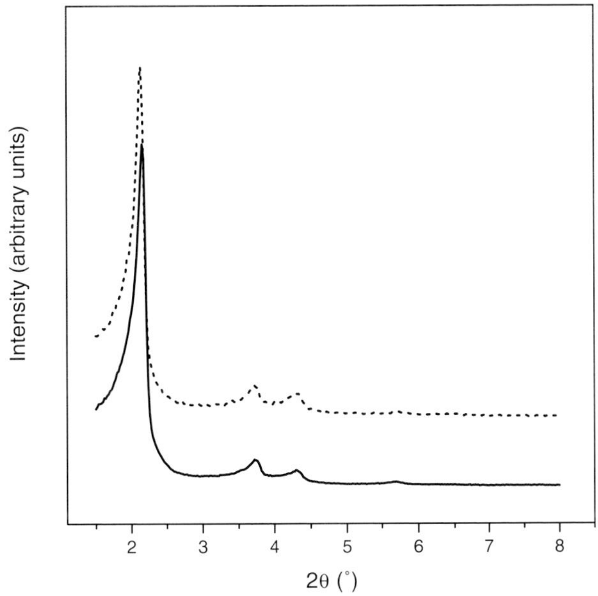

Figure 1 XRD patterns of the MCM-41 hexagonal ordered mesoporous solid (*solid line*) and the Fe_3O_4-loaded MCM-41 composite after iron loading and calcinations (*dashed lines*).

helium measurement has been performed yet on the γ-Fe_2O_3-loaded sample. The AC susceptibility measurements between 13 and 325 K were taken in a LakeShore 7,130 susceptometer using an exciting field amplitude of 1 Oe and a frequency of 825 Hz.

3 Results and discussion

The XRD patterns of the pure and the Fe_3O_4-loaded MCM-41 samples are displayed in Figure 1. The intense peak of the (100) reflection at $2\theta=2.1°$ is indicative that the structure of the hexagonal ordered mesoporous solid MCM-41 was preserved over the preparation and no change in its position was observed. The iron loading, determined by atomic absorption spectroscopy, was 7.7 wt.%.

Figure 2 shows the Mössbauer spectra the Fe_3O_4-loaded MCM-41 sample at the temperatures indicated. The room temperature spectrum is composed of a resolved magnetic signal (relative area ≈24%) and a central doublet. This spectrum was best-fitted to two sextets and a Fe^{3+} doublet, respectively. The hyperfine parameters obtained for the magnetic signal indicates that it belongs to Fe_3O_4 [16]. When the temperature decreases, the area of the magnetic signal increases and the background buckles. At 22 K a relaxing component also had to be included in the fitting. Finally, at 4.2 K the relative area of the doublet decreases down to 12% at the expense of the magnetic components.

The spectra at 298 and 22 K of the γ-Fe_2O_3-loaded sample are displayed in Figure 3. Both exhibit similar features to the magnetite-loaded sample. In this case, the quadrupole central signal at 298 K contributes with a relative area of 77% to the spectrum. The resolved magnetically blocked signal, which comprises the remaining spectrum area, can be fitted to only one hyperfine site whose parameters belong to γ-Fe_2O_3 [16]. This result confirms the complete oxidation of Fe_3O_4 to γ-Fe_2O_3 as a consequence of the thermal treatment in air. Very similar aspects of the spectrum can be observed at 22 K. At this temperature, all components that are in a relaxation regime at 298 K do not show any trend towards a

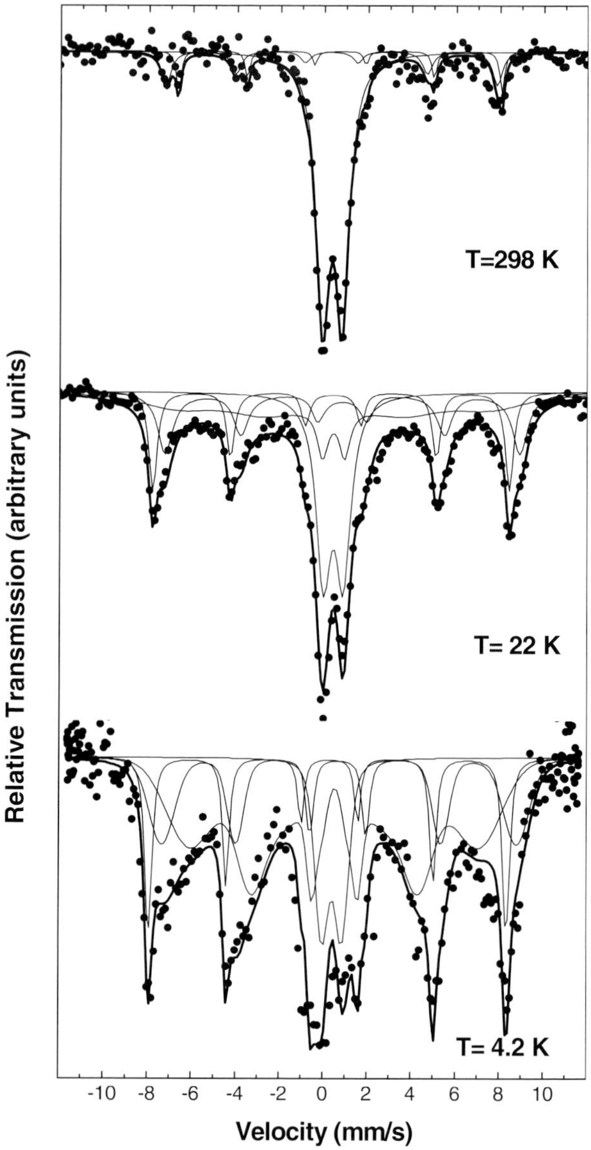

Figure 2 Mössbauer spectra at 298, 22 and 4.2 K of the Fe_3O_4-loaded MCM-41 sample. The *thick solid lines* are the results of the fitting procedure described in the text. The *thinner solid lines* are the components that make up the spectra.

splitting of their Mössbauer signal. This difference, in comparison with the Fe_3O_4-loaded MCM-41 sample, could be attributed to the existence of different magnetic anisotropy energy barriers between both compounds.

Figures 4 and 5 show the results of the susceptibility measurements. On decreasing the temperature the in-phase component of the Fe_3O_4-loaded sample, χ', monotonically decreases down to 110 K where a discontinuity is clearly observed. This change of slope corresponds to the Verwey transition temperature of Fe_3O_4, through which the net magnetic moment decreases because the antiferromagnetic exchange interaction between the hopping electrons and the inner electrons at iron cation octahedral sites becomes dominant when the

Figure 3 Mössbauer spectra at 298 and 22 K of the γ-Fe_2O_3-loaded MCM-41 sample. The *thick solid lines* are the results of the fitting procedure described in the text. The *thinner solid lines* are the components that make up the spectra.

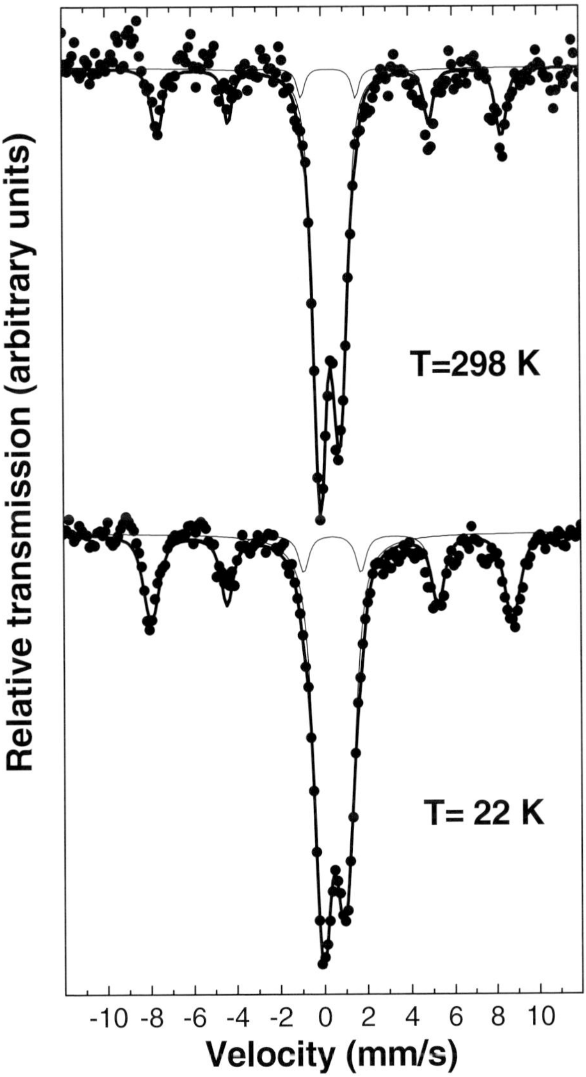

thermal energy decreases [17]. In addition, χ' shows a shoulder near 50 K. More clearly, the out-of-phase susceptibility component, χ'', directly associated to the magnetic energy dissipation, displays (Figure 5) a maximum at 42 K. An anomaly in χ' near 50 K has also been observed in microsized Fe_3O_4 which, although related to relaxation processes, is still not fully understood [17]. Zhang et al. [18] have observed a maximum at ≈50 K in the thermal dependence of the magnetization in Fe_3O_4 nanowires prepared by electro-deposition into anodic Alumina templates with pore diameters of 200 nm. This was interpreted as the Verwey transition of Fe_3O_4 taking place at a lower temperature than bulk Fe_3O_4 ($T_V \approx 110$–120 K) as a consequence of the reduced dimension of the array. A considerable shift of the Verwey transition temperature to lower values has also been observed for Fe_3O_4 nanoparticles of sizes ≈20 nm [19].

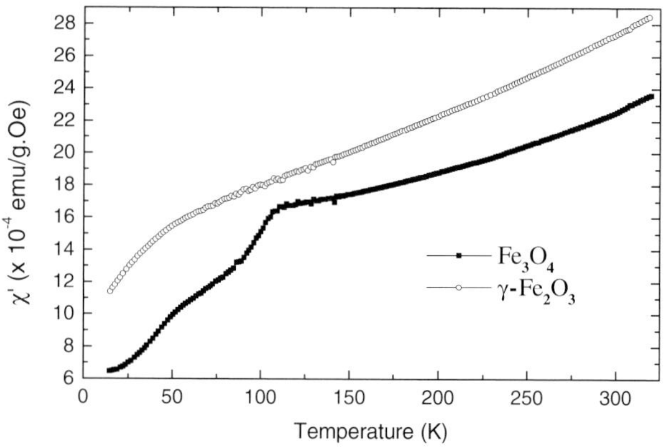

Figure 4 In-phase AC susceptibility (χ') measured with a frequency of 825 Hz and amplitude of 1 Oe. The *solid circles* belong to the Fe_3O_4-loaded sample and the *open* ones to the γ-Fe_2O_3-loaded sample.

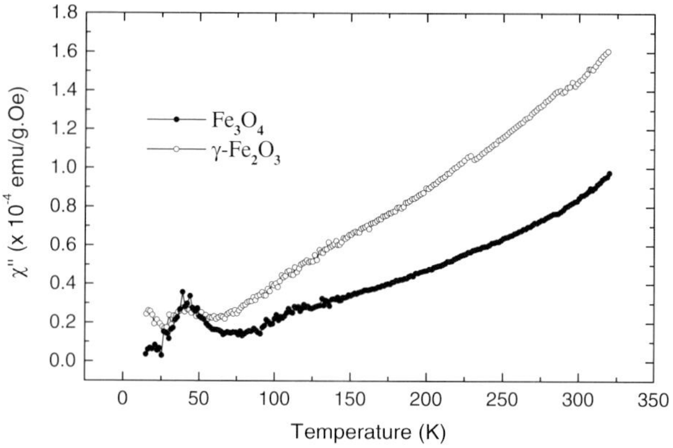

Figure 5 Out-of-phase AC susceptibility component (χ'') measured with a frequency of 825 Hz and amplitude of 1 Oe. The *solid circles* belong to the Fe_3O_4-loaded sample and the *open* ones to the γ-Fe_2O_3-loaded sample.

The thermal dependence of the susceptibilities χ' and χ'' of the γ-Fe_2O_3-loaded sample can also be seen in Figures 4 and 5. The Verwey transition is no longer observed but the shoulder at about 50 K is still apparent; as before, it is displayed more clearly in the out-of-phase component thermal dependence at a temperature of 41 K, which is coincident with the temperature observed for the magnetite-loaded sample.

In our case, those magnetite particles included within the MCM-41 channels could exhibit a shift of the Verwey transition, but, considering our Mössbauer and AC susceptibility results of the maghemite sample-that as discussed, have behaviors of $\chi'(T)$ and $\chi''(T)$ similar to the magnetite-loaded sample-we believe rather that the shoulder of χ' at 50 K and the χ'' maximum at $\approx$42 K are probably associated to the blocking of the magnetic moments of the fractions of smaller Fe_3O_4 and γ-Fe_2O_3 particles.

Finally, considering that surface effects are expected to take place because of the reduced pore diameters of the channels for those iron species embedded in the MCM-41 matrix, we assign the distributed signal at 4.2 K of the Fe_3O_4-loaded sample to Fe atoms located at the surface of these species. Their individual magnetic moments would be arranged in a partially randomly frozen way, as in a spin-glass-like state. On the other hand, the presence of a doublet at 4.2 K may indicate the existence of a paramagnetic species. Taking into account that the silylation treatment neutralizes the existing surface silanol groups, the assignment of this paramagnetic species to Fe^{3+} ions exchanged with the silicate framework [20] may be discarded. This signal might arise from isolated Fe-O-Fe entities covering the

MCM-41 pore walls; however, the current data do not display a noticeable paramagnetic-like behavior in the susceptibility at low temperatures.

A different explanation of the doublet at 4.2 K may come from a macroscopic quantum tunneling of the magnetization [21]. This assignment would require further experimental evidence before it can be stated unambiguously. We would need a thorough characterization of the sample, measurements over a range of temperatures that may separate the thermal activated regime from the macroscopic quantum tunneling, and experiments that can examine the predictions of the model based on reasonable parameters and assumptions regarding the magnetic particles. However, since one of the requirements of the theory is the existence of two very different magnetic anisotropy constants (perpendicular and parallel to the magnetic field) we may expect that if indeed fractions of the Fe species in our samples have arranged themselves forming quasi one-dimensional structures, the ratio of their longitudinal and transverse anisotropies can be very large.

Based on the above results, a possible interpretation of the synthesis process and the structure of the composite could be as follows. To produce a complete homogeneously hydrophobic surface that ensures an even iron species distribution within the MCM-41 channels, we neutralized the silanol groups with a silylation agent. This treatment would ensure that the iron species spread over the surface of the MCM-41 channels after impregnation. A major percentage of these species agglomerates into very fine particles that relax superparamagnetically at room temperature and progressively get blocked. The origin of the paramagnetic signal in the Mössbauer spectrum at 4.2 K of the magnetite MCM-41 loaded sample can arise from two causes. It may be caused by the fraction of very small particles that does not cluster and keeps on covering the inner surface in the form of a monolayer of Fe-O-Fe entities. An alternative explanation lies in the already mentioned macroscopic quantum tunneling of the magnetization. Both assignments would need further experimental work to be confirmed. The remaining percentage of iron oxides forms bigger particles or magnetic structures that display a blocked Mössbauer signal at room temperature.

4 Conclusions

Preliminary results on the impregnation of MCM-41 with an iron solution of $Fe(AcAc)_3$ in benzene after subjecting the matrix to a silylation treatment have successfully brought about Fe_3O_4- and γ-Fe_2O_3-loaded MCM-41 solids, where a significant fraction of the iron-oxide particles seem to be located inside the channels. Magnetic and hyperfine results allow concluding that both iron species are composed of a high percentage of small particles that relax superparamagnetically and a minor fraction of large particles in a blocked magnetic state. The paramagnetic signal apparent at 4.2 K in the Mössbauer spectrum of the sample of MCM-41 loaded with Fe_3O_4, may arise either from a small fraction of isolated Fe-O-Fe entities or from a macroscopic quantum tunneling of the magnetization.

References

1. Liu, S., Wang, Q., Van der Boort, P., Cool, P., Vansant, E.F., Jiang, M.: J. Magn. Magn. Mater. **280**, 31 (2004)
2. Jung, J.S., Choi, K.H., Chae, W.S., Kim, Y.R., Jun, J.H., Malkinski, L., Kodenkandath, T., Zhou, W., Wiley, J.B., O'Connor, C.J.: J. Phys. Chem. Solids **64**, 385 (2003)
3. Pankhurst,Q.A., Pollard, R.J.: J. Phys., Condens. Matter **5**, 8487 (1993)
4. Zhang, L., Papaefthymiou, G.C., Ziolo, R.F., Ying, J.Y.: Nanostruct. Mater. **9**, 185 (1997)

5. Xue, D.S., Zhang, L.Y., Gao, C.X., Xu, X.F., Gui, A.B.: Chin. Phys. Lett. **21**, 733 (2004)
6. Zhang, L., Papaefthymiou, G.C., Ying, J.Y.: J. Appl. Phys. **81**, 6892 (1997)
7. Zhang, L.Y., Xue, D.S., Xu, X.F., Gui, A.B.: J. Magn. Magn. Mater. **294**, 10 (2005)
8. Zhang, L., Papaefthymiou, G.C., Ying, J.Y.: J. Phys. Chem., B. **105**, 7414 (2001)
9. Morales, A.M., Lieber, C.M.: Science. **279**, 208 (1998)
10. Wu, C.W., Aoki, T., Kuwabara, M.: Nanotechnology. **15**, 1886 (2004)
11. Crowley, T.A., Ziegler, K.J., Lyons, D.M., Erts, D., Olin, H., Morris, M.A., Holmes, J.D.: Chem. Mater. **15**, 3518 (2003)
12. Cagnoli, M.V., Alvarez, A.M., Bengoa, J.F., Gallegos, N.G., Mogni, L.V., González Oddera, M.X., Marchetti, S.G., Mercader, R.C.: Hyperfine Interact. **148**, 185 (2003)
13. Bengoa, J.F., Cagnoli, M.V., Gallegos, N.G., Alvarez, A.M., Mogni, L.V., Moreno, M.S., Marchetti, S. G.: Microporous Mesoporous Mater. **84**, 153 (2005)
14. Cagnoli, M.V., Gallegos, N.G., Bengoa, J.F., Alvarez, A.M., Moreno, M.S., Roig, A., Marchetti, S.G., Mercader, R.C.: American Institute of Physics Conference Proceedings **765**, 13 (2005)
15. Ryoo, R., Kim, J.M.: J. Chem. Soc. Chem. Commun. **8**, 711–712 (1995)
16. Vandenberghe, R.E., De Grave, E., Landuydt, C., Bowen, L.H.: Hyperfine Interact. **53**, 175 (1990)
17. Muxworthy, A.R., McClelland, E.: Geophys. J. Int. **140**, 101 (2000)
18. Zhang, L.Y., Xue, D.S., Xu, X.F., Gui, A.B., Gao, C.X.: J. Phys., Condens. Matter **16**, 4541 (2004)
19. Goya, G.F., Berquo, T.S., Fonseca, F.C., Morales, M.P.: J. Appl. Phys. **94**, 3520 (2003)
20. Selvam, P., Dapurkar, S.E., Badamali, S.K., Murugasan, M., Kuwano, H.: Catal. Today **68**, 69 (2001)
21. Gunther, L. In: Dormann, J.L., Fiorani, D. (eds.) Magnetic properties of fine particles, p. 213. North Holland, Amsterdam (1992)

Hyperfine Interact (2006) 170:83–89
DOI 10.1007/s10751-006-9470-7

Nuclear forward scattering vs. conventional Mössbauer studies of atomically tailored Eu-based materials

A. Konjhodzic · A. Adamczyk · F. Vagizov · Z. Hasan · E. E. Alp · W. Sturhahn · Jiyong Zhao · J. J. Carroll

Published online: 9 January 2007

Abstract With the decrease in size of devices, rapid characterization of nano-devices is an inevitable necessity. It is shown that Mössbauer spectroscopy using synchrotron radiation from the advanced photon source provides such a tool of investigation. Results are presented and compared for conventional Mössbauer and Nuclear Forward Scattering for ^{151}Eu-doped magnesium sulfide as an example, especially at low concentrations.

Key words europium · nuclear forward scattering · Mössbauer spectroscopy · atomic tailoring · photonic materials · rare-earths

1 Introduction

Mössbauer spectroscopy is an extremely powerful technique of characterization for samples rich in iron and a few other isotopes. Using this method, different valence states of these isotopes as well as small variations in the crystalline field environment around them in a solid can be studied systematically and rapidly. As the size of the sample becomes microscopic, or the concentration of Mössbauer active nuclei decreases, the technique loses its effectiveness. In such cases, at least a lengthy data accumulation to enhance the signal to noise ratio becomes compulsory. However, for some experiments, the data accumulation time becomes prohibitively long. In this category, there are experiments performed on samples with extremely low concentrations of Mössbauer active ions or the ones that show extremely small effects resulting in the signal being buried in the noise. Nuclear Forward Scattering (NFS) using X-rays or γ-rays from a synchrotron source offers an alternative to

A. Konjhodzic · A. Adamczyk · F. Vagizov · Z. Hasan (✉)
Physics Department, Temple University, Philadelphia, PA 19122, USA
e-mail: zhasan@temple.edu

E. E. Alp · W. Sturhahn · J. Zhao
Advanced Photon Source, Argonne National Laboratory, Argonne, IL 60439, USA

J. J. Carroll
Department of Physics & Astronomy, Youngstown State University, Youngstown, OH 44555, USA

the conventional Mössbauer technique [1–3]. The radiation from the synchrotron source is intense, tunable in energy, and in the form of short pulses to provide the opportunity of time domain γ-ray spectroscopy.

This paper deals with the application of Mössbauer spectroscopy in identifying samples doped with Eu and its different valence states in extremely low concentrations in alkaline earth sulfides. We discuss the case of MgS in particular. Low concentrations of impurities are necessary for photonics applications where the impurity atoms or ions should be isolated from each other. These isolated atoms/ions have well defined energy levels that can be used for the desired applications. At high concentrations, however, pair effects arising from Coulomb or exchange interactions can significantly change the energy level structure from a single ion to a more complicated one. In general, these pairs are difficult to deal with theoretically and unpredictable for any practical use.

Typical impurity concentrations for photonics applications range around 0.01 mol% while the size of the Photonic devices can be as small as a few cubic microns. Using conventional Mössbauer spectroscopy, weeks or months of data accumulation time would be needed for such samples. This is partly because strong ^{151}Sm sources are not readily available. Therefore, synchrotron radiation offers a great opportunity to study the Mössbauer spectrum by nuclear forward scattering. A comparison of the Mössbauer spectroscopy between using a conventional ^{151}Sm source and the synchrotron radiation source is presented below.

2 Atomically tailored photonics materials

The size of photonic devices is always decreasing and in many cases has reached the quantum regime. In principle, the ultimate lower limit on the size of a photonic device is a single atom. However, to be useful this atom should be localized and optically addressable. Such is the case with optically active impurities in a wide bandgap semiconductor or an insulator. Therefore, in the future, only a few selected optically active atoms or ions in a solid will form the ‘core’ of the quantum photonic devices. Some fascinating applications that could work with only a few atoms in a solid host are: (1) spectral storage, where the information is stored in the optical spectrum of a small number of selected atoms in a solid, (2) quantum computing with a few optically addressable atoms, and, (3) coherent control of nuclear states with a laser [4–7]. In all these and similar atomic scale applications, there are severe constraints imposed on the energy level structure of the host, the optically active ion, its nucleus, and even the nuclei of the host material.

In spectral storage, for example, the electronic states should be tailored to strongly absorb the photons from a laser, the photons that are used for addressing or photo-transforming the ions. The energy of the optical transition should be tailored to be preferably in the range of commercial lasers. The electronic states should be suitably located in the bandgap of the host to enhance (or in some cases to suppress) two-photon ionization and other non-linear optical processes. And finally, the relative concentration of different ionization states of the active ion should be controlled to facilitate a particular photo-induced transformation.

For spectral storage, MgS and CaS were prepared with a very low concentration of Eu, 0.01 mol%. In these europium based II–VI sulfides, the information is stored in the spectrum of Eu^{2+} by photoionizing them with a laser. A relatively large concentration of Eu^{3+} is necessary for trapping the electrons liberated in the process of photoionization.

Therefore, doubly and triply positive valence states and the electronic energy levels of Eu were tailored to meet the requirements of an ideal material for spectral storage. It is not possible to determine the relative concentration of Eu^{2+}/Eu^{3+} in MgS using optical techniques. This is particularly true at low Eu concentrations. Eu^{3+} has weakly allowed optical transitions within the states of $4f^6$ configuration. Coincidently, very strongly allowed 4f–5d transitions of Eu^{2+} also lie around the same energy range in this host. The two spectra overlap considerably and any quantitative determination of relative Eu^{2+}/Eu^{3+} concentrations is not possible. However, ^{151}Eu (nat. abun. ~48%) is a Mössbauer isotope. Therefore, this technique can provide the information about the Eu^{2+}/Eu^{3+} concentration ratio.

Samples of MgS doped with varying concentrations of Eu were prepared by the methods described elsewhere [8, 9]. The laser-excited fluorescence spectra at ~15 K gave strong $4f^7$–$4f^6 5d^1$ emission confirming the presence of Eu^{2+} in the sample. In MgS, Eu^{2+} substitutes for Mg^{2+} and therefore it is an energetically more favored ionization state of Eu. Eu^{3+} can also occupy a substitutional site with charge compensation provided by long-range displacement of the lattice ions. This displacement maintains the cubic site symmetry and, therefore, both Eu^{2+} and Eu^{3+} occupy sites of octahedral symmetry [10]. A slight variation from the cubic symmetry is there, which is expected due to the size mismatch between Eu^{2+}/Eu^{3+} and Mg^{2+}, the ion for which they substitute in the lattice (Ionic radii: 1.17, 0.95 and 0.72 Å, respectively, for Eu^{2+}, Eu^{3+} and Mg^{2+}). This relaxes the cubic selection rules for the 4f–5d electronic transition resulting in very strong Zero Phonon Lines (ZPL) in the optical absorption spectrum; an attribute of great importance for quantum photonic applications.

3 Mössbauer spectroscopy of ^{151}Eu in MgS

Mössbauer spectra of MgS:Eu samples, with naturally abundant Eu, (^{151}Eu:^{153}Eu, 0.478:0.522), were taken using a $^{151}SmF_3$ source. Figure 1 shows the spectrum of $Mg_{0.8}Eu_{0.2}S$ at room temperature. The data accumulation was done in the constant velocity mode. It lasted for 14 h to give a signal of 8% absorption. The absorption spectrum shows a peak at 11.7 mm/s corresponding to $^{151}Eu^{2+}$. A small shoulder is also clearly visible on the high velocity side, v=13.9 mm/s. The data has been fitted to two Lorentzian lines and has been resolved in the diagram. Due to the broad linewidth of the SmF_3 source, ~2 mm/s, this feature cannot be further resolved, even at low temperature. The origin of this small peak has not been investigated in detail here. However, such features could arise from non-standard Eu^{2+} – centers that could be formed due to the trace impurities. Our sample, MgS, is particularly prone to such impurities – it readily absorbs water from the environment and is known to form oxygen-associated Eu centers. Also, prolonged Mössbauer experiments expose the sample to such elements more easily. The sample is in the form of polycrystalline powder where the surface to volume ratio is high and small traces of water can be absorbed readily.

The spectrum in Fig. 1 does not show any trace of Eu^{3+} resonance which in this host should appear at ~0 mm/s. This clearly indicates that almost all Eu, 95(±5)% is in the form of Eu^{2+}. The uncertainty of ±5% is due to the poor signal to noise ratio of the absorption. The line-fitting of the spectrum in Fig. 1 shows that 88(±1)% of Eu^{2+} is at the normal substitutional site and the remaining 12(±1)% is at a site associated with some impurity. In our case, oxygen usually does not appear as an impurity. The appearance of this shoulder, therefore, could be attributed to prolonged exposure of the sample to the environment. Optical investigations on this sample are underway in our group.

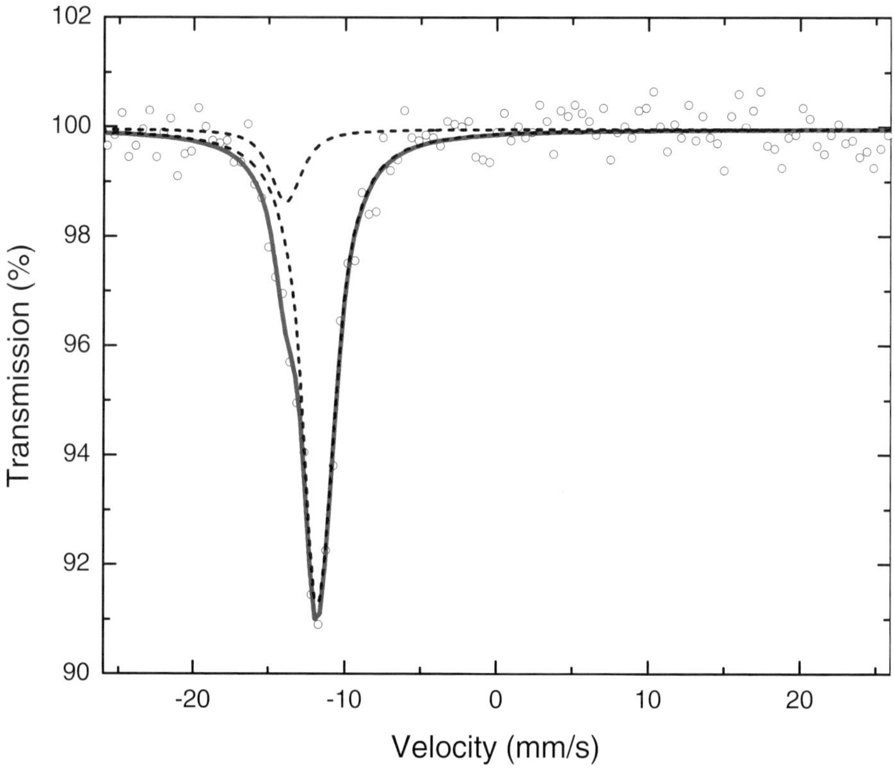

Fig. 1 Mössbauer spectrum of MgS:Eu (20%), data acquisition time 14 h. The data can be fitted to *two lines*, arising from two different Eu^{2+} centers as shown by the *dotted lines* (Isomer shifts: 13.9 and 11.7 mm/s, linewidths: 2.4 and 2.2 mm/s). The relative contribution of these two centers is 88% and 12%

Figure 2 shows the Mössbauer spectrum of MgS with 0.01 mol% of Eu. The spectrum was taken under exactly the same conditions as the experiment of Fig. 1. However, there were two exceptions: (1) the data accumulation was carried out for a period of 6 days. (2) the sample thickness was increased to 2 mm as opposed to ~0.5 mm for 20% Eu sample. The Eu^{2+} resonance at −12.6 mm/s is ~5% absorption above the noise level and the Eu^{3+} signal is clearly visible at +0.5 mm/s. Within the resolution of the absorption lines, the two lines appear as isolated single lines. By comparing the area under the two resonances, the ratio of Eu^{2+}/ Eu^{3+} is ~0.65/0.35 in this sample. A single line for Eu^{2+} further means that for this sample there is only one dominant Eu^{2+} site, i.e., the substitutional cubic site with Eu^{2+} replacing Mg^{2+}. The broadening of Mössbauer resonances could also be due to the temperature effects and quadrupole splittings of Eu^{3+} that are in the tens of MHz range. However, these effects are beyond the resolution of our SmF_3 source.

4 Nuclear forward scattering of MgS:Eu

Mössbauer studies, using the synchrotron at the Advanced Photon Source (APS), Argonne National Laboratory were performed on the same samples as used in Figs. 1 and 2.

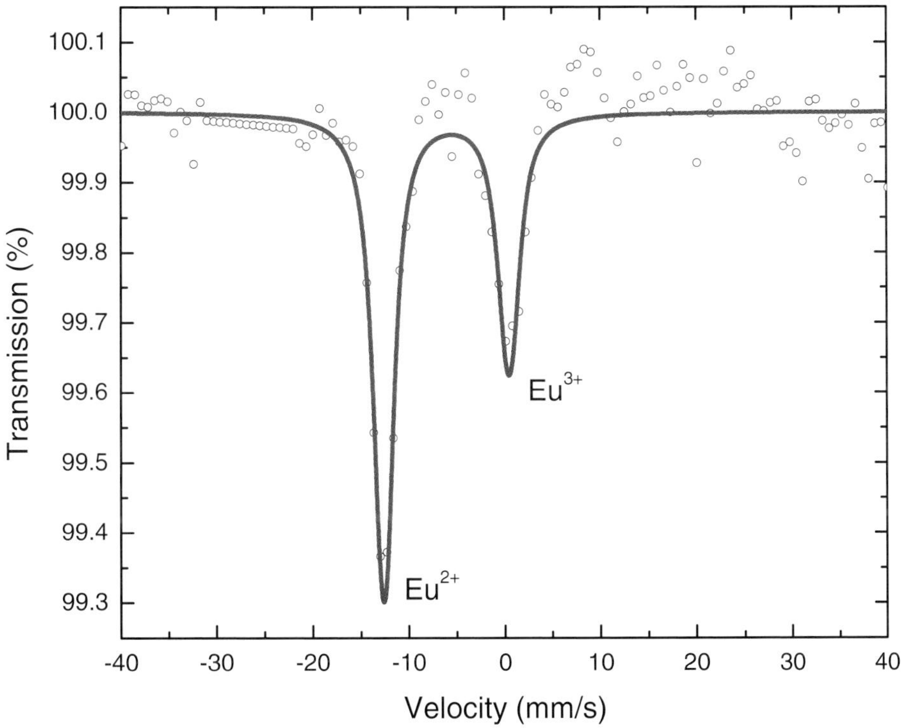

Fig. 2 Mössbauer spectrum of MgS:Eu (0.01 mol%) using a $^{151}SmF_3$ source, data acquisition time 6 days. Isomer shifts for Eu^{2+} and Eu^{3+} signals are, respectively, −12.6 and +0.5 mm/s. The corresponding linewidths are 2.4 and 2.6 mm/s, respectively

Figure 3 shows the nuclear forward scattering data on 20% Eu sample. The details of the experiments and the characteristics of the beam are described elsewhere [11]. To observe the beating of Eu^{3+} and Eu^{2+} signals over the time, a known amount of Eu_2O_3 was mixed in MgS:Eu (20 %) sample which predominantly had europium as Eu^{2+}. The CONUSS program [12] was used for fitting the decay of the beating signal. The fitting shown as the dashed curve in Fig. 3 used the relative Isomer Shift (IS=−12.8 mm/s), quadrupole splitting for Eu^{3+} (2.1 mm/s) and the relative concentration of Eu^{2+}: Eu^{3+} (0.35:0.65) from the known value from the data of Fig. 1. Only small variations, within the errors of determining these parameters were found to give the best fit. The isomer shift used in Fig. 3 is different from that used in Fig. 1, as we have not considered two different Eu^{2+} centers present in the sample as was clearly seen in Fig. 1. Secondly, in NFS (Fig. 3) we see relative isomer shift between Eu^{2+} in MgS and Eu^{3+} in Eu_2O_3 rather than SmF_3 (Fig. 1).

The total data accumulation time was 20 min for the spectrum of Fig. 3. This direct comparison between conventional Mössbauer and Nuclear Forward Scattering gives a ratio of 24. Roughly this is a factor by which the time of experiment is reduced from conventional Mössbauer to the Mössbauer studies using the time domain Nuclear Forward Scattering. In MgS samples with 0.01 mol% Eu the NFS experiment required a data collection time of about 6 h to give a good beating signal between the Eu^{3+} and Eu^{2+} resonances as opposed to 6 days of collection time for the conventional Mössbauer spectra of Fig. 2. Once again this is roughly a reduction by a factor of 24. However, note in these estimates, the ratio of the sizes of the samples is not taken into consideration. NFS is a very

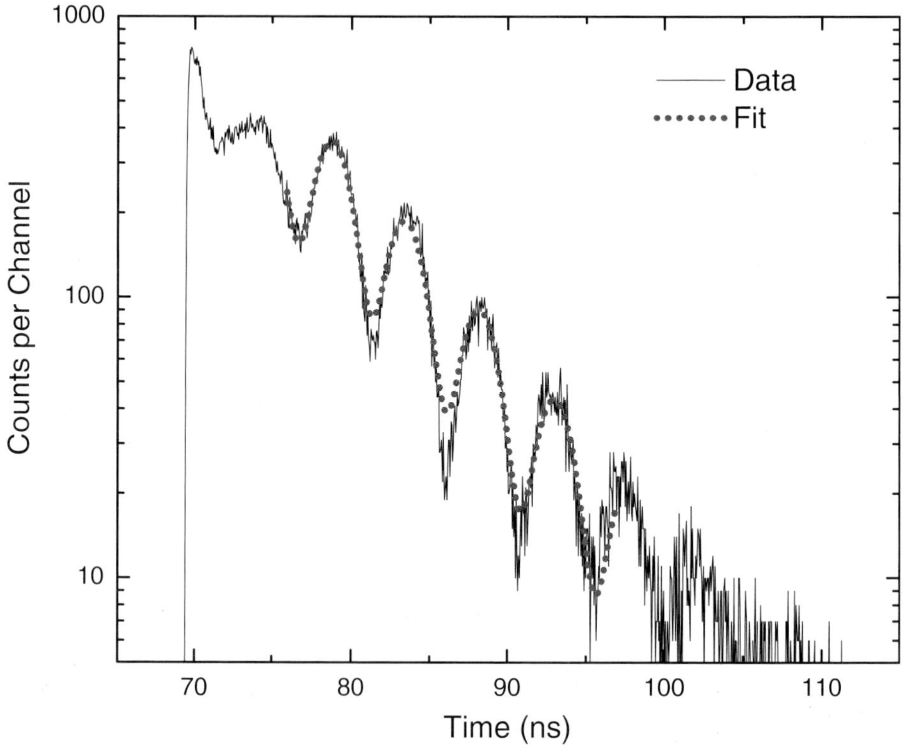

Fig. 3 Nuclear forward scattering using 21.54 keV radiation for ^{151}Eu from the APS. The sample was 70.7 μm thick and was composed of MgS:Eu (20%) mixed with Eu_2O_3 for observing the beating of the Eu^{3+} and Eu^{2+} signals. The relative concentration Eu^{3+}:Eu^{2+} was 0.65:0.35 and the data was accumulated for 20 min. The fitting of the decay with CONUSS program yielded relative isomer shift of −12.8 mm/s and the quadrupole splitting of 2.1 mm/s

sensitive technique and the sample thicknesses used were ~50–100 μm, smaller by an order of magnitude than what were used for the conventional Mössbauer experiments. If this is taken into consideration, then by using NFS assisted Mössbauer, the overall data collection times are reduced by more than two orders of magnitude. Thus, thin samples such as films, or samples with a very small concentration of Mössbauer isotopes are at a greater advantage with regard to the expediency of data accumulation in NFS assisted Mössbauer.

5 Conclusions

In conclusion, we have shown that Nuclear Forward Scattering using fast pulses from a synchrotron source has great promise for fast characterization. In MgS:Eu, the valence states of Eu and their electronic states have been tailored for extremely low concentration samples that are necessary for some photonic applications. We have presented and compared data on Mössbauer studies by the conventional technique and using nuclear forward scattering in this material. The main advantage of NFS is that it reduces the data collection times by orders of magnitude when compared to the conventional Mössbauer technique. This is hoped to enable quick characterization studies on samples that have extremely small amounts of Mössbauer isotopes. Even more excitingly, with such reduction in data accumulation times it is hoped that some very novel experiments can be performed where small effects are supposed to give very weak observable signals. Such weak signals necessitate prohibitively long data acquisition times. One of these experiments is the laser control of nuclear states. In this experiment a strong laser is expected to pump the electronic state and by electron-nuclear coupling, the nucleus of the parent Eu ion can be coherently controlled that can be detected by Mössbauer spectroscopy. It is estimated that for a detectable signal months of data accumulation time may be required using conventional Mössbauer technique. NFS has potential to reduce this time to hours or even minutes.

Acknowlegment This work was made possible by grants from Temple University Office of Research, the Office of Naval Research (grant no. N00014-00-1-0371, N00014-02-1-0869), AFOSR (grant no. 49620-01-1-0279, 49620-96-1-0347) and DARPA (grant no. F49620-01-1-0566). We greatly appreciate the help from Mike Campanell in preparing the manuscript.

References

1. Leupold, O., Pollmann, J., Gerdau, E., Rüter, H.D., Faigel, G., Tegze, M., Bortel, G., Rüffer, R., Chumakov, A.I., Baron, A.Q.R.: Europhys. Lett. **35**, 671 (1996)
2. Koyama, I., Yoda, Y., Zhang, X.W., Ando, M., Kikuta, S.: Jpn. J. Appl. Phys. **35**, 6297 (1996)
3. Leupold, O., Chumakov, A.I., Alp, E.E., Sturhahn W., Baron, A.Q.R.: Hyperfine Interact. **123–124**, 611 (1999)
4. Hasan, Z., Mathur, V.K.: U.S. Patent Numbers: 6,528,234 (2003)
5. Mathur, V.K., Hasan, Z.: U.S. Patent Numbers: 6,514,435 (2003)
6. Basieva, I.T., Sekatskii, S.K., Pukhov, K.K., Basiev, T.T., Dietler, G.: Laser Phys. **14**, 1393 (2004)
7. Roman Kolesov, Yuri Rostovstev, Olga Kocharovskaya.: Opt. Commun. **179**, 537 (2000)
8. Hasan, Z., Biyikli, L., Macfarlane, P.I.: Appl. Phys. Lett. **72**, 3399 (1998)
9. Hasan, Z., Solonenko, M., Macfarlane, P.I., Biyikli, L., Mathur, V.K., Karwacki, F.A.: Appl. Phys. Lett. **72**, 2373 (1998)
10. Pandey, R., Sivaram, S.: J. Phys. Chem. Solids **52**, 211 (1991)
11. Toellner, T.S., Hu, M.Y., Sturhahn, W., Quast, K., Alp, E.E.: Appl. Phys. Lett. **71**, 15 (1997)
12. Sturhahn, W., Gerdau, E.: Phys. Rev. B **49**, 9285 (1994)

Hyperfine Interact (2006) 170:91–104
DOI 10.1007/s10751-006-9467-2

A complete solution to the Mössbauer problem, all in one place

C. J. Voyer · D. H. Ryan

Published online: 21 December 2006

Abstract We present a full solution to the general combined interactions static Mössbauer problem that is easily generalized to any Mössbauer isotope, and applies for M1, E1, and E2 transitions as well as combined M1–E2 transitions. Explicit expressions are given for both powder and single crystal samples.

Key words Mössbauer spectrum calculation · full Hamiltonian diagonalization · combined hyperfine interactions

1 Introduction

While first-order perturbation approaches or interpolation schemes are adequate for fitting the majority of simple Mössbauer spectra, there are numerous cases for which these approaches are invalid. If contributions from magnetic and electrostatic interactions are of comparable magnitude or do not share a common quantization axis, a full solution to the nuclear Hamiltonian must be used to fit the spectrum.

Although functional codes have been published in the past [1–3], in the current computational climate, where not only programming languages compete, but where computional packages that include programmable options (Matlab, Mathematica…) are also used, an explicit and adaptable mathematical solution is more useful. Most calculations given are either not worked out completely [1, 3], or are far too involved mathematically to be easily generalized and implemented into simple programs [4].

In order to promote the more widespread implementation of full solution Mössbauer fitting codes, we present here a complete solution that we have developed. This solution can be generalized to cover any accessible Mössbauer isotope and is cast in a form that can be readily coded into a fitting routine.

C. J. Voyer · D. H. Ryan (✉)
Centre for the Physics of Materials and Physics Department, McGill University, 3600 University Street, Montréal, QC H3A 2T8, Canada
e-mail: dominic@physics.mcgill.ca

2 The Mössbauer problem

The Mössbauer problem consists of computing a Mössbauer spectrum given a set of hyperfine parameters. The solution can then be used in a least squares refinement of an experimentally obtained Mössbauer spectrum. The line positions depend on the splitting of the ground and excited states and their relative energy difference, and can be derived from the eigenvalues of the system Hamiltonian. Line intensities depend on the coupling of the two angular momentum states involved in the transition, and can be obtained from the eigenvectors of the Hamiltonian combined with appropriate vector coupling coefficients [5]. In the absence of hyperfine fields, the ground and excited states are degenerate, and a single line spectrum is observed. When an electric field gradient, or a magnetic field, or both are present, the degeneracy is lifted and multi-line spectra are obtained as shown in Figure 1.

In the static model, the interaction Hamiltonian $\mathcal{H}$ is given by [6]:

$$\mathcal{H} = \mathcal{H}_\circ + \mathcal{H}_C + \mathcal{H}_M + \mathcal{H}_Q + \ldots \tag{1}$$

$\mathcal{H}_\circ$ represents all terms for the atom not including hyperfine interactions and does not influence the energy difference between the ground and excited states of the nucleus. Having no impact on the Mössbauer spectrum, it is ignored. $\mathcal{H}_C$ represents the Coulombic interactions between the nucleus and the electrons, i.e. the electric monopole term, and has the effect of shifting all excited state energy levels with respect to all ground state energy levels, and is called the isomer shift (IS). The IS will have the effect of shifting the entire Mössbauer spectrum either up or down in velocity. Other effects can also shift the entire spectrum by some constant velocity value, and so for fitting purposes we consider the more general centre shift (CS), which can be added to any solution once the more complex calculations are completed. Finally, $\mathcal{H}_M$ and $\mathcal{H}_Q$, the magnetic dipole and electric quadrupole contributions, are the terms that lift the energy level degeneracy, and are the non-trivial part of the computation. All higher order terms are ignored [6].

2.1 Magnetic dipole interactions

For a nucleus in a state with a gyromagnetic ratio g, and subject to a hyperfine magnetic field $\mathbf{B}$, $\mathcal{H}_M$ can written as [7]:

$$\mathcal{H}_M = -g\mu_N \mathbf{B} \cdot \hat{\mathbf{I}} \tag{2}$$

where $\hat{\mathbf{I}}$ is the total spin operator and μ_N is the nuclear magneton. For simplicity, we may choose the z-axis to be the direction of the field and get:

$$\mathcal{H}_M = -g\mu_N B \hat{I}_z \tag{3}$$

where B is the magnitude of the field and $\hat{I}_z$ is the z component of the angular momentum operator. For both the excited and ground states, the $|m\rangle$ states are eigenstates of this Hamiltonian and thus the eigenvalues are:

$$E_M = -g\mu_N Bm \tag{4}$$

and the eigenvector matrix is simply identity.

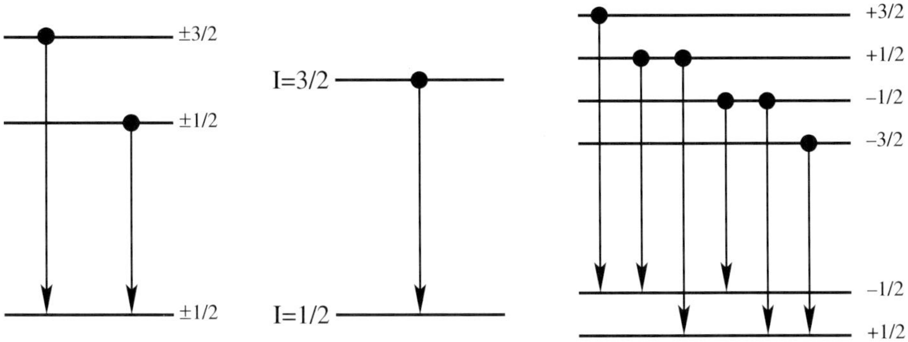

Figure 1 Sketch of energy level degeneracy lifting in the ^{57}Fe case. In the *centre* we see the degenerate excited and ground states, on the *left* the excited state has the degeneracy partially lifted when an electric field gradient is present, and on the *right* when a magnetic field is present, we see the degeneracy completely lifted. The *arrows drawn* are the allowed transitions given the M1 multipolarity.

2.2 Electric quadrupole interactions

The nuclear energy level degeneracy is also lifted by the quadrupole moment of the probe nucleus interacting with an electric field gradient. The electric field gradient is a 3×3 traceless tensor:

$$V_{ij} = (\partial^2 V / \partial x_i \partial x_j) \qquad (x_i, x_j = x, y, z). \tag{5}$$

The coordinate system can always be chosen such that the tensor is diagonal and traditionally the z-axis is chosen so that V_{zz} is its largest component. Since the tensor is traceless, $V_{xx} + V_{yy} + V_{zz} = 0$, only two parameters are required to specify the tensor completely. We define $\eta = \frac{V_{xx} - V_{yy}}{V_{zz}}$, using the convention $|V_{zz}| \geq |V_{yy}| \geq |V_{xx}|$ to ensure that $0 \leq \eta \leq 1$, and the tensor can be specified by V_{zz} and η . For a nucleus with an electric quadrupole moment Q, spin I, subject to an electric field gradient whose principal axis value is V_{zz}, with an asymmetry η, the Hamiltonian is [8]:

$$\mathcal{H}_Q = \frac{eQV_{zz}}{4I(2I-1)} \left[3\hat{I}_z^2 - I(I+1) + \frac{\eta}{2}(\hat{I}_+^2 - \hat{I}_-^2) \right] \tag{6}$$

where e is the charge of the proton and $\hat{I}_+$ and $\hat{I}_-$ are the ladder operators as defined by Sakurai [9].

If $\eta = 0$ the $|m\rangle$ states are energy eigenstates and the energies are $E_m = \frac{eQV_{zz}}{4I(2I-1)} \left[3m^2 - I(I+1) \right]$, which in turn yields a degeneracy in the energy levels as only the magnitude of m contributes to the energy. Complications arise when $\eta \neq 0$ since in this case the $|m\rangle$ states are no longer energy eigenstates. The Hamiltonian is then non-diagonal and the eigenvectors need to be computed explicitly and a full diagonalization is necessary.

A fitting routine needs to compute a full spectrum many times in order to complete a pattern refinement. Available computing power used to limit solution possibilities, and strategies were devised to avoid repeated calculation of the full solution. Exact analytic solutions for the energy levels exist for the $I = \frac{3}{2}$ case [10], and the $I = \frac{5}{2}$ case

[11], however, when the dimensions of the Hamiltonian get larger the mathematics of an analytic solution becomes prohibitively difficult. Shenoy et al. [12] generated a polynomial interpolation scheme, where the line positions and intensities have been computed exactly for incremental values of η and then fit with a polynomial whose coefficients were made available. Finally the perturbative approach is useful if one of the interactions is small compared to the other. Today, however, computing power limitations for this type of calculation no longer exist, and the only remaining barrier to a full solution is the mathematics involved.

The ideal situation is to have a solution to the Mössbauer problem that can be implemented easily as a numerical routine that yields exact line positions and intensities given a set of hyperfine parameters, with no limitations on the values of these parameters. This solution could then be incorporated into a least squares fitting routine and refine experimentally obtained data. Numerical diagonalization is longer a limitation. We can solve the general static Mössbauer problem exactly, and have a solution that can cover the most complex cases, as well as covering simple cases, that can be easily implemented, operate efficiently and satisfy all fitting needs.

3 Combined interactions

The most general physical situation that can be considered is the case of combined magnetic dipole and electric quadrupole interactions. In the previous cases considered, the optimal choice of quantization axis was evident. Here we have two logical choices: the magnetic field direction, or the principal axis of the electric field gradient, V_{zz}. Typically, V_{zz} is chosen as the quantization axis and the schematic setup is shown in Figure 2. Note that this choice of axes is valid even if $V_{zz} = 0$, and none of the computations change.

From (1), the Hamiltonian operator for this system becomes [8]:

$$\mathcal{H} = \mathcal{H}_Q + \mathcal{H}_M \tag{7}$$

where $\mathcal{H}_Q$ does not change from (6), but $\mathcal{H}_M$ needs to be rotated away from the z-axis and can be written following the convention of Figure 2 as:

$$\mathcal{H}_M = -g\mu_N B_{hf}\left[\hat{I}_z\cos\theta + \frac{1}{2}(\hat{I}_+ e^{-i\phi} + \hat{I}_- e^{i\phi})\sin\theta\right] \tag{8}$$

This Hamiltonian will not, in general, be diagonal in the $|m\rangle$ basis.

To compute the energy eigenvalues and eigenvectors we need to write the Hamiltonian in matrix form. The expression for the matrix depends on the total spins I_g and I_e of the ground and excited states. Once the Mössbauer transition is chosen, and I_g and I_e are thereby fixed, we can write the Hamiltonian for the ground and excited states, $\mathcal{H}_g$ and $\mathcal{H}_e$.

For example, if we are interested in a $\frac{3}{2} \rightarrow \frac{1}{2}$ transition (as in ^{57}Fe, ^{119}Sn ...) or a $\frac{1}{2} \rightarrow \frac{3}{2}$ (as in ^{189}Os, ^{197}Au...) transition, if we set $A = \frac{eQV_{zz}}{4I(2I-1)}$ where Q is the appropriate value for the quadrupole moment of the nucleus in the $I = \frac{3}{2}$ state, and

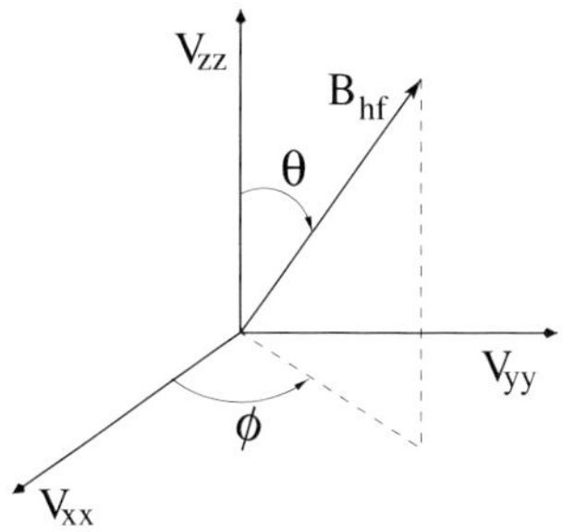

Figure 2 Schematic of the ϕ and θ angles, V_{zz} and B_{hf} as they appear in the Hamiltonian of (7).

$\alpha = g_{\frac{3}{2}} \mu_N B_{hf}$, where $g_{\frac{3}{2}}$ is the appropriate value for the gyromagnetic factor of the nucleus in the $I = \frac{3}{2}$ state, we may write (matrices constructed as described in [13]):

$$\mathcal{H}_{3/2} = \begin{pmatrix} 3A - \frac{3}{2}\alpha\cos\theta & -\frac{\sqrt{3}}{2}\alpha\sin\theta e^{-i\phi} & \sqrt{3}A\eta & 0 \\ -\frac{\sqrt{3}}{2}\alpha\sin\theta e^{i\phi} & -3A - \frac{\alpha}{2}\cos\theta & -\alpha\sin\theta e^{-i\phi} & \sqrt{3}A\eta \\ \sqrt{3}A\eta & -\alpha\sin\theta e^{i\phi} & -3A + \frac{\alpha}{2}\cos\theta & -\frac{\sqrt{3}}{2}\alpha\sin\theta e^{-i\phi} \\ 0 & \sqrt{3}A\eta & -\frac{\sqrt{3}}{2}\alpha\sin\theta e^{i\phi} & 3A + \frac{3}{2}\alpha\cos\theta \end{pmatrix} \tag{9}$$

and setting $\beta = g_{1/2}\mu_N B_{hf}$:

$$\mathcal{H}_{1/2} = \begin{pmatrix} -\frac{\beta}{2}\cos\theta & -\frac{\beta}{2}\sin\theta e^{-i\phi} \\ -\frac{\beta}{2}\sin\theta e^{+i\phi} & +\frac{\beta}{2}\cos\theta \end{pmatrix} \tag{10}$$

The $I = 0$ case is of particular mathematical interest because the Hamiltonian in (7) is null. This case is treated by considering the state to have a unidimensional eigenvector in the $|m\rangle$ basis, and an energy eigenvalue of 0. So if we are interested in the $2 \rightarrow 0$ transition (^{170}Yb, ^{166}Er...), the only Hamiltonian that needs to be explicitly computed is the $I = 2$ excited state Hamiltonian (here $\beta = g_2\mu_N B_{hf}$):

$$\mathcal{H}_2 = \begin{pmatrix} 6A - 2\beta\cos\theta & -\beta e^{-i\phi}\sin\theta & A\sqrt{6}\eta & 0 & 0 \\ -\beta e^{i\phi}\sin\theta & -3A - \beta\cos\theta & -\frac{\sqrt{6}}{2}\beta\sin\theta e^{-i\phi} & 3A\eta & 0 \\ A\sqrt{6}\eta & -\beta\frac{\sqrt{6}}{2}e^{i\phi}\sin\theta & -6A & -\frac{\sqrt{6}}{2}\beta\sin\theta e^{-i\phi} & B\sqrt{6}\eta \\ 0 & 3A\eta & -\frac{\sqrt{6}}{2}\beta\sin\theta e^{i\phi} & -3A + \beta\cos\theta & -\beta\sin\theta e^{-i\phi} \\ 0 & 0 & A\sqrt{6}\eta & -\beta\sin\theta e^{i\phi} & 6A + 2\beta\cos\theta \end{pmatrix} \tag{11}$$

With the excited and ground state Hamiltonians now analytically expressed, a routine may compute the components, given a set of hyperfine parameters, then proceed to the numerical diagonalization, and finally compute the line positions and intensities.

Many options are available for the numerical diagonalization. We have used the `tred2` and `tqli` routines outlined in Numerical Recipes [14, 15]. The Matlab and Mathematica software packages also offer diagonalization functions. To avoid working with complex numbers Numerical Recipes suggests a scheme by which the matrix can be kept real, however it requires doubling the dimensionality of the problem. After diagonalization is achieved, the problem is then reduced back to its original dimension by selecting half of the eigenvalues and eigenvectors. The selection must be done with care since the ordering of the diagonalization output depends on the set of non-zero hyperfine parameters. The selection must ensure a set of mutually orthogonal eigenvectors is obtained.

From these routines, we obtain numerical values for our eigenvectors $|e_n\rangle$ as well as the amplitudes $\langle n|m\rangle$ and the eigenvalues of the system. For a state with total spin I and corresponding eigenvalues e_n, for example:

$$|e_n\rangle = \begin{pmatrix} \langle m = I|n_{e_n}\rangle \\ \langle m = I-1|n_{e_n}\rangle \\ \langle m = I-2|n_{e_n}\rangle \\ \vdots \end{pmatrix} \tag{12}$$

At this stage we have the eigenvalues which will give us the line positions, and the $\langle m|n\rangle$ factors which will give us the intensity of each line. We have thus obtained all the necessary information from the Hamiltonians of the excited and ground states, and may now proceed to calculate the exact line positions and intensities.

3.1 Line position calculation

The energy shift detected by Mössbauer spectroscopy is the difference between the energies of the excited state Hamiltonian E_{n_e} and the ground state Hamiltonian E_{n_g}. For a given transition between the levels E_{n_g} of the ground state and E_{n_e} of the excited state, we obtain a line position $E_n(e \to g)$:

$$E_n(e \to g) = E_{n_e} - E_{n_g} \tag{13}$$

The constant offset CS mentioned in Section 2 can then be added, and the final result is completely general for all values of the hyperfine parameters. We note that for the specific case of the $2 \to 0$ transition (^{170}Yb), this reduces to:

$$E_n(2 \to 0) = E_{n_2} + CS \tag{14}$$

The number of lines observed will depend on the dimensionality of both $\mathcal{H}_e$ and $\mathcal{H}_g$, and also on the multipolarity of the transition i.e. the quantum mechanical selection rules that apply in each situation.

3.2 Line intensity calculation

The relative probability of each transition's occurrence will depend on the coupling of the two angular momentum states involved in the transition [5]. Several methods exist for calculating the relative intensity of a line [3, 5, 8, 16]. These treatments fall short however, in that they either treat a single case [8], do not treat the E2 case [16], do not give an explicit and easily generalizable solution [3, 5, 8, 16], or do not cover the single crystal case [3, 5, 8].

We chose to treat the relative intensity of a line in a spectrum as the normalized power absorbed by the corresponding transition. We write the multipole electric field vector **E** as [17]:

$$\mathbf{E}(\varphi, \vartheta) = \sum_{l,m} (-i)^{l+1} \left[a_E(l, m)\mathbf{X}_l^m \times \mathbf{r} + a_M(l, m)\mathbf{X}_l^m \right] \tag{15}$$

where ϑ and φ are the polar angles describing the direction of emission of the γ, along **r**, with respect to our chosen axis of quantization (not to be confused with ϕ and θ defined in Section 3); $\mathbf{X}_l^m$ are the vector spherical harmonics, and will be derived later (Section 3.3); and the $a_E(l, m)$ and $a_M(l, m)$ specify the amounts of electric multipole and magnetic multipole fields, and are determined by the source and by boundary conditions (derived in Section 3.4) [18].

Since we are only concerned with relative intensities, factors common to all transitions can be normalized out and so are omitted from (15). The power absorbed (i.e. the intensity of a line) will be $\mathbf{E}.\mathbf{E}^*$. This method simply and naturally introduces the vector spherical harmonics needed, and leads to a straightforward single crystal intensity as well as a powder average intensity. This method is preferred because of its simple computations and adaptability to different transitions the experimentalist may face.

The three transition multipolarities encountered in Mössbauer spectroscopy are the M1, E1 and E2 cases. Only the M1 and E2 are ever combined, and the E1 case is completely analogous to the M1 case and requires no new information, thus for the purposes of this computation, we will not address the E1 case.

From (15) we get for the M1 case,

$$\mathbf{E}_{M1}(\varphi, \vartheta) = - \sum_{m=-1}^{+1} a_M(1, m).\mathbf{X}_l^m \tag{16}$$

and for the E2 case we get

$$\mathbf{E}_{E2}(\varphi, \vartheta) = \sum_{m=-2}^{+2} i.\left[a_E(2, m)\right].(\mathbf{X}_l^m \times \mathbf{r}) \tag{17}$$

To proceed further, we need to express **E** more explicitly, thus calculating the vector spherical harmonics $\mathbf{X}_l^m$, and the amounts of electric and magnetic multipole fields $a_E(l, m)$ and $a_M(l, m)$.

3.3 The $\mathbf{X}_l^m$

The vector spherical harmonics, $\mathbf{X}_l^m$, can be generated from [18]

$$\mathbf{X}_l^m = \frac{1}{\sqrt{l(l+1)}} \hat{\mathbf{L}} Y_l^m(\vartheta, \varphi) \tag{18}$$

where,

$$\hat{\mathbf{L}} = i\left(\hat{\vartheta}\frac{1}{\sin\vartheta}\frac{\partial}{\partial\varphi} - \hat{\varphi}\frac{\partial}{\partial\vartheta}\right) \tag{19}$$

We compute the vector spherical harmonics needed,

$$\mathbf{X}_1^{-1} = \frac{1}{4}\sqrt{\frac{3}{\pi}}\left(e^{-i\varphi}\hat{\vartheta} - i\cos\vartheta e^{-i\varphi}\hat{\varphi}\right) \tag{20}$$

$$\mathbf{X}_1^{0} = \frac{1}{4}\sqrt{\frac{3}{\pi}}i\sqrt{2}\sin\vartheta\hat{\varphi} \tag{21}$$

$$\mathbf{X}_1^{1} = \frac{1}{4}\sqrt{\frac{3}{\pi}}\left(e^{i\varphi}\hat{\vartheta} + i\cos\vartheta e^{i\varphi}\hat{\varphi}\right) \tag{22}$$

$$\mathbf{X}_2^{-2} = \frac{1}{4}\sqrt{\frac{5}{\pi}}\left(\sin\vartheta e^{-2i\varphi}\hat{\vartheta} - i\sin\vartheta\cos\vartheta e^{-2i\varphi}\hat{\varphi}\right) \tag{23}$$

$$\mathbf{X}_2^{-1} = \frac{1}{4}\sqrt{\frac{5}{\pi}}\left(\cos\vartheta e^{-i\varphi}\hat{\vartheta} - i\left[\cos^2\vartheta - \sin^2\vartheta\right]e^{-i\varphi}\hat{\varphi}\right) \tag{24}$$

$$\mathbf{X}_2^{0} = \frac{1}{4}\sqrt{\frac{5}{\pi}}\left(\sqrt{6}i\cos\vartheta\sin\vartheta\right)\hat{\varphi} \tag{25}$$

$$\mathbf{X}_2^{1} = \frac{1}{4}\sqrt{\frac{5}{\pi}}\left(\cos\vartheta e^{i\varphi}\hat{\vartheta} + i\left[\cos^2\vartheta - \sin^2\vartheta\right]e^{i\varphi}\hat{\varphi}\right) \tag{26}$$

$$\mathbf{X}_2^{2} = \frac{1}{4}\sqrt{\frac{5}{\pi}}\left(-\sin\vartheta e^{2i\varphi}\hat{\vartheta} - i\sin\vartheta\cos\vartheta e^{2i\varphi}\hat{\varphi}\right) \tag{27}$$

We note that in spherical coordinates $\hat{\vartheta}\times\hat{r} = \hat{\varphi}$ and that $\hat{\varphi}\times\hat{r} = -\hat{\vartheta}$, such that for the $\mathbf{X}_2^m$ of the E2 part we get

$$\mathbf{X}_2^{-2}\times\hat{r} = \frac{1}{4}\sqrt{\frac{5}{\pi}}\left(\sin\vartheta e^{-2i\varphi}\hat{\varphi} + i\sin\vartheta\cos\vartheta e^{-2i\varphi}\hat{\vartheta}\right) \tag{28}$$

$$\mathbf{X}_2^{-1}\times\hat{r} = \frac{1}{4}\sqrt{\frac{5}{\pi}}\left(\cos\vartheta e^{-i\varphi}\hat{\varphi} + i\left[\cos^2\vartheta - \sin^2\vartheta\right]e^{-i\varphi}\hat{\vartheta}\right) \tag{29}$$

$$\mathbf{X}_2^{0}\times\hat{r} = \frac{1}{4}\sqrt{\frac{5}{\pi}}\left(-\sqrt{6}i\cos\vartheta\sin\vartheta\right)\hat{\vartheta} \tag{30}$$

$$\mathbf{X}_2^{1}\times\hat{r} = \frac{1}{4}\sqrt{\frac{5}{\pi}}\left(\cos\vartheta e^{i\varphi}\hat{\varphi} - i\left[\cos^2\vartheta - \sin^2\vartheta\right]e^{i\varphi}\hat{\vartheta}\right) \tag{31}$$

$$\mathbf{X}_2^{2}\times\hat{r} = \frac{1}{4}\sqrt{\frac{5}{\pi}}\left(-\sin\vartheta e^{2i\varphi}\hat{\varphi} + i\sin\vartheta\cos\vartheta e^{2i\varphi}\hat{\vartheta}\right) \tag{32}$$

3.4 The $a(\mathbf{L}, m)$

For this particular problem, the $a_E(\mathbf{L}, m)$ and $a_M(\mathbf{L}, m)$ coefficients have the same analytic expression [16],

$$a(\mathbf{L}, m) = \sum_{m_e - m_g = m} \langle m_e | n_e \rangle \langle m_g | n_g \rangle^* \langle I_g \mathbf{L} m_g m | I_e m_e \rangle \tag{33}$$

where $|n_g\rangle$ and $|n_e\rangle$ are the energy eigenstates of the ground and excited states respectively, and $|m_e\rangle$ and $|m_g\rangle$ are the $\hat{I}_z$ eigenstates for the excited and ground states respectively, (note that $\langle m_e|n_e\rangle$ and $\langle m_g|n_g\rangle$ are simply the components of the eigenvectors of $\mathcal{H}_e$ and $\mathcal{H}_g$, respectively), and $\langle I_g \mathbf{L} m_g m | I_e m_e \rangle$ are Clebsch-Gordan coefficients where $\mathbf{L}$ is the multipolarity of the transition (for M1, $\mathbf{L} = 1$ and for E2, $\mathbf{L} = 2$), $m = m_e - m_g$, I_e is the total spin of the excited state and I_g is the total spin of the ground state.

For the $2 \rightarrow 0$ transition, all Clebsch-Gordon coefficients are exactly 1 since $I_g = 0$. But if the spins for both the excited and ground states are non-zero, the Clebsch-Gordon coefficients are non-trivial (see Table I for an example). Note that the $\langle I_g \mathbf{L} m_g m | I_e m_e \rangle$ coefficients are first converted using the relation [19]:

$$\langle I_g \mathbf{L} m_g m | I_e m_e \rangle = (-1)^{I_g - m_g} \sqrt{\frac{2I_e + 1}{2(I_e - I_g) + 1}} \langle I_e I_g m_e - m_g | \mathbf{L} m \rangle \tag{34}$$

Once an expression for the Clebsch-Gordon coefficients is obtained we can move on to calculating the $a(\mathbf{L}, m)$ explicitly. Again, for the case of a $3/2 \rightarrow 1/2$ transition we get,

$$a(2,2) = -\sqrt{\frac{4}{5}} \langle 3/2|n_e\rangle \langle -1/2|n_g\rangle^*$$

$$a(2,1) = \sqrt{\frac{1}{5}} \langle 3/2|n_e\rangle \langle 1/2|n_g\rangle^* - \sqrt{\frac{3}{5}} \langle 1/2|n_e\rangle \langle -1/2|n_g\rangle^*$$

$$a(2,0) = \sqrt{\frac{2}{5}} \langle 1/2|n_e\rangle \langle 1/2|n_g\rangle^* - \sqrt{\frac{2}{5}} \langle -1/2|n_e\rangle \langle -1/2|n_g\rangle^*$$

$$a(2,-1) = \sqrt{\frac{3}{5}} \langle -1/2|n_e\rangle \langle 1/2|n_g\rangle^* - \sqrt{\frac{1}{5}} \langle -3/2|n_e\rangle \langle -1/2|n_g\rangle^*$$

$$a(2,-2) = \sqrt{\frac{4}{5}} \langle -3/2|n_e\rangle \langle 1/2|n_g\rangle^*$$

$$a(1,1) = \sqrt{\frac{1}{3}} \langle 1/2|n_e\rangle \langle -1/2|n_g\rangle^* + \langle 3/2|n_e\rangle \langle 1/2|n_g\rangle^*$$

$$a(1,0) = \sqrt{\frac{2}{3}} \langle 1/2|n_e\rangle \langle 1/2|n_g\rangle^* + \sqrt{\frac{2}{3}} \langle -1/2|n_e\rangle \langle -1/2|n_g\rangle^*$$

$$a(1,-1) = \sqrt{\frac{1}{3}} \langle -1/2|n_e\rangle \langle 1/2|n_g\rangle^* + \langle -3/2|n_e\rangle \langle -1/2|n_g\rangle^*$$

Table I Clebsch-Gordon coefficients pertinent to the $3/2 \rightarrow 1/2$ M1 and/or E2 solution

m_g	m_e	m	E2 $\langle \frac{1}{2}\mathbf{2}m_g m \vert \frac{3}{2}m_e\rangle$	M1 $\langle \frac{1}{2}\mathbf{1}m_g m \vert \frac{3}{2}m_e\rangle$
$+\frac{1}{2}$	$+\frac{3}{2}$	$+1$	$\sqrt{\frac{1}{5}}$	1
$+\frac{1}{2}$	$+\frac{1}{2}$	0	$\sqrt{\frac{2}{5}}$	$\sqrt{\frac{2}{3}}$
$+\frac{1}{2}$	$-\frac{1}{2}$	-1	$\sqrt{\frac{3}{5}}$	$\sqrt{\frac{1}{3}}$
$+\frac{1}{2}$	$-\frac{3}{2}$	-2	$\sqrt{\frac{4}{5}}$	0
$-\frac{1}{2}$	$+\frac{3}{2}$	$+2$	$-\sqrt{\frac{4}{5}}$	0
$-\frac{1}{2}$	$+\frac{1}{2}$	$+1$	$-\sqrt{\frac{3}{5}}$	$\sqrt{\frac{1}{3}}$
$-\frac{1}{2}$	$-\frac{1}{2}$	0	$-\sqrt{\frac{2}{5}}$	$\sqrt{\frac{2}{3}}$
$-\frac{1}{2}$	$-\frac{3}{2}$	-1	$-\sqrt{\frac{1}{5}}$	1

where the $a(2, m)$ are for the E2 multipolarity case, and the $a(1, m)$ are for the M1 multipolarity case. For the $2 \rightarrow 0$ case with E2 multipolarity however, the expressions become much simpler,

$$
\begin{aligned}
a(2, 2) &= \langle 2|n_e\rangle \\
a(2, 1) &= \langle 1|n_e\rangle \\
a(2, 0) &= \langle 0|n_e\rangle \\
a(2, -1) &= \langle -1|n_e\rangle \\
a(2, -2) &= \langle -2|n_e\rangle
\end{aligned}
$$

The vectors in (16) and (17) now have numerical values. We can now move on to calculating the intensity of each line.

3.5 M1 transitions

We first consider the case of an M1 transition (^{57}Fe, ^{119}Sn...). The electric field vector can be written as:

$$
\mathbf{E}_{M1}(\varphi, \vartheta) = \frac{1}{4}\sqrt{\frac{5}{\pi}}\sqrt{\frac{3}{5}}\Big(\Big[a(1,1)e^{i\varphi} + a(1,-1)e^{-i\varphi}\Big]\hat{\vartheta} + i\Big[a(1,1)e^{i\varphi}\cos\vartheta + a(1,0)\sqrt{2}\sin\vartheta - a(1,-1)e^{-i\varphi}\cos\vartheta\Big]\hat{\varphi}\Big) \tag{35}
$$

Noting that $\hat{\varphi}.\hat{\vartheta} = 0$, the intensity of a line can be written as:

$$\begin{aligned}\mathbf{E}_{M1}.\mathbf{E}^*_{M1} = \frac{3}{16\pi}\Big(& |a(1,1)|^2(1+\cos^2\vartheta) + 2|a(1,0)|^2\sin^2\vartheta \\ & + |a(1,-1)|^2(1+\cos^2\vartheta) + 2\Re\left[a(1,1)a^*(1,-1)e^{2i\varphi}\right] \\ & + 2\Re\left[a(1,1)a^*(1,0)\sqrt{2}\sin\vartheta\cos\vartheta e^{i\varphi}\right] \\ & - 2\Re\left[a(1,1)a^*(1,-1)\cos^2\vartheta e^{2i\varphi}\right] \\ & - 2\Re\left[a(1,0)a^*(1,-1)\sqrt{2}\sin\vartheta\cos\vartheta e^{i\varphi}\right]\Big)\end{aligned} \tag{36}$$

Equation 36 is the intensity for a line in a single crystal Mössbauer absorption with a γ absorption that makes angles φ and ϑ with respect to the principal axis of the electric field gradient V_{zz}. The $\frac{3}{16\pi}$ factor can be ignored as it is the same for all transitions.

Most samples used are polycrystalline, in which the intensity needs to be integrated over all directions. The intensity for a powder sample becomes:

$$\int_{\vartheta=0}^{\pi}\int_{\varphi=0}^{2\pi} \mathbf{E}_{M1}.\mathbf{E}^*_{M1}\sin\vartheta\, d\varphi d\vartheta = |a(1,1)|^2 + |a(1,0)|^2 + |a(1,-1)|^2 \tag{37}$$

We note that although one of the most cited solutions of the Mössbauer problem (Kündig's solution of the ^{57}Fe problem [8]) yields a slightly different expression, his expression can be simplified to obtain (37), providing an independent verification that our process is correct.

3.6 E2 transitions

For an E2 transition the selection rules are less restrictive and far more transitions are allowed. The electric field vector can be written as:

$$\begin{aligned}\mathbf{E}_{E2} = \frac{1}{4}\sqrt{\frac{5}{\pi}}\Big(\Big[& -a(2,-2)\sin\vartheta\cos\vartheta e^{-2i\varphi} - a(2,-1)\left[\cos^2\vartheta - \sin^2\vartheta\right]e^{-i\varphi} \\ & + a(2,0)\sqrt{6}\cos\vartheta\sin\vartheta + a(2,1)\left[\cos^2\vartheta - \sin^2\vartheta\right]e^{i\varphi} \\ & - a(2,2)\sin\vartheta\cos\vartheta e^{2i\varphi}\Big]\hat{\vartheta} + i\Big[a(2,-2)\sin\vartheta e^{-2i\varphi} \\ & + a(2,-1)\cos\vartheta e^{-i\varphi} + a(2,1)\cos\vartheta e^{i\varphi} - a(2,2)\sin\vartheta e^{2i\varphi}\Big]\hat{\varphi}\Big)\end{aligned} \tag{38}$$

The intensity is then:

$$\begin{aligned}\mathbf{E}_{E2}.\mathbf{E}^*_{E2} = \frac{5}{16\pi}\Big(& |a(2,-2)|^2 \sin^2\vartheta(1+\cos^2\vartheta) + |a(2,-1)|^2\left(\left[\cos^2\vartheta - \sin^2\vartheta\right]^2 + \cos^2\vartheta\right)\\
& + 6|a(2,0)|^2\cos^2\vartheta\sin^2\vartheta + |a(2,1)|^2\left(\left[\cos^2\vartheta - \sin^2\vartheta\right]^2 + \cos^2\vartheta\right)\\
& + |a(2,2)|^2\left(\sin^2\vartheta\cos^2\vartheta + \sin^2\vartheta\right)\\
& + 2\Re\left\{a(2,-2)a^*(2,-1)\sin\vartheta\cos\vartheta\left[\cos^2\vartheta - \sin^2\vartheta\right]e^{-i\varphi}\right\}\\
& - 2\Re\left\{a(2,-2)a^*(2,0)\sqrt{6}\cos^2\vartheta\sin^2\vartheta e^{-2i\varphi}\right\}\\
& - 2\Re\left\{a(2,-2)a^*(2,1)\sin\vartheta\cos\vartheta\left[\cos^2\vartheta - \sin^2\vartheta\right]e^{-3i\varphi}\right\}\\
& + 2\Re\left\{a(2,-2)a^*(2,2)\cos^2\vartheta\sin^2\vartheta e^{-4i\varphi}\right\}\\
& - 2\Re\left\{a(2,-1)a^*(2,0)\sqrt{6}\sin\vartheta\cos\vartheta\left[\cos^2\vartheta - \sin^2\vartheta\right]e^{-i\varphi}\right\}\\
& - 2\Re\left\{a(2,-1)a^*(2,1)\left[\cos^2\vartheta - \sin^2\vartheta\right]^2 e^{-2i\varphi}\right\}\\
& + 2\Re\left\{a(2,-1)a^*(2,2)\sin\vartheta\cos\vartheta\left[\cos^2\vartheta - \sin^2\vartheta\right]e^{-3i\varphi}\right\}\\
& + 2\Re\left\{a(2,0)a^*(2,1)\sqrt{6}\sin\vartheta\cos\vartheta\left[\cos^2\vartheta - \sin^2\vartheta\right]e^{-i\varphi}\right\}\\
& - 2\Re\left\{a(2,0)a^*(2,2)\sqrt{6}\cos^2\vartheta\sin^2\vartheta e^{-2i\varphi}\right\}\\
& - 2\Re\left\{a(2,1)a^*(2,2)\sin\vartheta\cos\vartheta\left[\cos^2\vartheta - \sin^2\vartheta\right]e^{-i\varphi}\right\}\\
& + 2\Re\left[a(2,-2)a^*(2,-1)\cos\vartheta\sin\vartheta e^{-i\varphi}\right]\\
& + 2\Re\left[a(2,-2)a^*(2,1)\sin\vartheta\cos\vartheta e^{-3i\varphi}\right]\\
& - 2\Re\left[a(2,-2)a^*(2,2)\sin^2\vartheta e^{-4i\varphi}\right]\\
& + 2\Re\left[a(2,-1)a^*(2,1)\cos^2\vartheta e^{-2i\varphi}\right]\\
& - 2\Re\left[a(2,-1)a^*(2,2)\sin\vartheta\cos\vartheta e^{-3i\varphi}\right]\\
& - 2\Re\left[a(2,1)a^*(2,2)\cos\vartheta\sin\vartheta e^{-i\varphi}\right]\Big) \qquad (39)\end{aligned}$$

Again, as was done in Section 3.5, we integrate over all directions to obtain the intensities for the powder average case and obtain,

$$\int_{\vartheta=0}^{\pi}\int_{\varphi=0}^{2\pi}\mathbf{E}.\mathbf{E}^*\sin\vartheta\,\mathrm{d}\vartheta\,\mathrm{d}\varphi = |a(2,-2)|^2 + |a(2,-1)|^2 + |a(2,0)|^2 + |a(2,1)|^2 + |a(2,2)|^2 \qquad (40)$$

a result which is analogous to the M1 case and completes the list of results necessary to compute the exact line intensity of any static combined interactions Mössbauer spectrum.

For the $2 \to 0$ case, the intensity expression is greatly simplified:

$$\begin{aligned}\int_{\vartheta=0}^{\pi}\int_{\varphi=0}^{2\pi} \mathbf{E}.\mathbf{E}^{*} \sin\vartheta \, d\vartheta \, d\varphi &= |\langle 2|i\rangle|^2 + |\langle 1|i\rangle|^2 + |\langle 0|i\rangle|^2 + |\langle -1|i\rangle|^2 + |\langle -2|i\rangle|^2 \\ &= \langle i| \left(\sum_m |m\rangle \langle m| \right) |i\rangle \\ &= 1 \end{aligned} \tag{41}$$

making the intensity computation unnecessary.

3.7 Mixed M1–E2 transitions

Certain Mössbauer isotopes have a mixture of M1 and E2 transitions in non-negligible amounts (e.g. ^{197}Au where E2/M1=0.11). If we define $\mathcal{X} = \frac{E2}{E2+M1}$, the intensity expression is:

$$\mathbf{E}.\mathbf{E}^{*} = (1 - \mathcal{X})\, \mathbf{E}_{M1}.\mathbf{E}^{*}_{M1} + \mathcal{X}\mathbf{E}_{E2}.\mathbf{E}^{*}_{E2} \tag{42}$$

Thus the single crystal intensity for a combined E2-M1 transition is the weighted sum of (36) and (39), and the powder intensity is the weighted sum of (37) and (40).

4 Adapting the solution to other isotopes

When faced with an isotope not treated in the present text, the researcher may simply apply the same general strategy used here, and indeed the more complex expressions still hold. (36), (37), (39) and (40) do not change.

A new isotope brings new values for the total spin of the excited and ground states and different values for the quadrupole moments and gyromagnetic ratios for the excited and ground states. The spectra obtained from different isotopes will behave differently but the strategies outlined in this paper can be easily applied:

1. Set up the matrices $\mathcal{H}_e$ and $\mathcal{H}_g$, as was done in (9), (10) and (11).
2. Set up the chosen diagonalization routine and store proper set of eigenvalues and eigenvectors
3. Calculate the sum in (33) using the appropriate Clebsch-Gordon coefficients. The number of terms is set by: $m_e - m_g = m$, and the $\langle m|n\rangle$ are the components of the eigenvectors as shown in (12).
4. The intensities are computed using the relevant equation from (36),(37), (39), (40), or (42), depending on the transition multipolarity (M1,E2, or M1–E2) and whether the experiment is on a single crystal or powder sample.
5. The line positions are calculated using (13).

This strategy is completely general for any static combined interactions case, for any Mössbauer isotope.

References

1. Gabriel, J.R., Ruby, S.L.: Nucl. Instrum. Methods **36**, 23 (1965)
2. Kündig, W.: Nucl. Instrum. Methods **75**, 336 (1969)
3. Chipaux, R.: Comput. Phys. Commun. **60**, 405 (1990)
4. Szymański, K.: J. Phys., Condens. Matter **12**, 7495 (2000)
5. Gibb, T.C., Greenwood, N.N.: In: Mössbauer Spectroscopy, p.66. Chapman and Hall, London, UK (1971)
6. Gibb, T.C., Greenwood, N.N.: In: Mössbauer Spectroscopy, p.46. Chapman and Hall, London, UK (1971)
7. Gibb, T.C., Greenwood, N.N.: In: Mössbauer Spectroscopy, p.60. Chapman and Hall, London, UK (1971)
8. Kündig, W.: Nucl. Instrum. Methods **48**, 219 (1967)
9. Sakurai, J.J.: In: Modern Quantum Mechanics, 2nd edn., p.192. Addison-Wesley, New York (1994)
10. Gibb, T.C., Greenwood, N.N.: In: Mössbauer Spectroscopy, p.56. Chapman and Hall, London, UK (1971)
11. Gerdau, E., Wolf, J., Winkler, H., Braunsfurth, J.: Proc. Roy. Soc. Ser. A **311**, 197–206 (1969)
12. Shenoy, G.K., Dunlap, B.D.: Nucl. Instrum. Methods **71**, 285 (1969)
13. Sakurai, J.J.: In: Modern Quantum Mechanics, 2nd edn., p.20. Addison-Wesley, New York (1994)
14. Press, W.H., Tukolsky, S.A., Vetterling, W.T., Flannery, B.P.: In: Numerical Recipes in C, 2nd edn., p.474. Cambridge University Press, New York (1999)
15. Press, W.H., Tukolsky, S.A., Vetterling, W.T., Flannery, B.P.: In: Numerical Recipes in C, 2nd edn., p.480. Cambridge University Press, New York (1999)
16. Housley, R.M., Grant, R.W., Gonser, U.: Phys. Rev. Lett. **178**, 514 (1969)
17. Jackson, J.D.: In: Classical Electrodynamics, 3rd edn., p.437. Wiley, Hoboken, NJ (1999)
18. Jackson, J.D.: In: Classical Electrodynamics, 3rd edn., p.431. Wiley, Hoboken, NJ (1999)
19. Edmonds, A.R.: In: Angular Momentum in Quantum Mechanics, p.42. Princeton University Press, Princeton, New Jersey (1957)

Hyperfine Interact (2006) 170:105–116
DOI 10.1007/s10751-006-9468-1

Anisotropic contributions to the transferred hyperfine field studied using a field-induced spin-reorientation

Laura K. Perry · D. H. Ryan · G. Venturini

Published online: 9 January 2007

Abstract We report here a comparison between a field-driven spin-flop ($TbMn_6Sn_{5.46}In_{0.54}$) and a temperature-driven spin reorientation ($TbMn_6Sn_{6-x}Ga_x$) in order to demonstrate that the anisotropic contribution to B_{hf} at the Sn sites can be obtained through the moment reorientation and is independent of the driving force. We show that a complete 90° spin reorientation can be achieved at 300 K in an applied field of 0.57(3) T and that the changes in hyperfine field due to the anisotropic contribution exceed 45% at one of the Sn sites. Quantitative values for the anisotropic constant at the three Sn sites are obtained.

Key words field-driven spin-flop · temperature-driven spin reorientation · moment reorientation · anisotropic contribution

1 Introduction

As Sn is non-magnetic, any hyperfine field measured at a Sn site in a crystal structure must be transferred to it from surrounding magnetic moments. This transferred hyperfine field has two main contributions [1]: isotropic and anisotropic. The isotropic contribution is due to conduction band polarization, which induces a spin imbalance at the nucleus. This is commonly referred to as the Fermi contact field, and depends only on the magnitude of the neighbouring moments and the symmetry of their magnetic structure. The anisotropic contribution involves the bonding between the

L. K. Perry · D. H. Ryan (✉)
Center for the Physics of Materials, Department of Physics, McGill University, Rutherford Building, 3600 University Street, H3A 2T8 Montréal, Québec, Canada
e-mail: dhryan@physics.mcgill.ca

L. K. Perry · G. Venturini
Laboratoire de Chimie du Solide Minérale, Université Henri Poincaré Nancy I, UMR 7555 BP 239, 54506 Vandoeuvre-les-Nancy, France

magnetic ion and the non-magnetic Mössbauer probe, leading to a transferred field which depends not only on the magnitude of the neighbouring moments, but also on the relative orientations of those moments and their connecting bonds. In the tetragonal $CuAl_2$-type TSn_2 (T = Mn, Fe) stannides, the isotropic contribution to the transferred hyperfine field at some ($MnSn_2$ [2]) or all ($FeSn_2$ [3]) of the Sn sites cancels, resulting in a transferred hyperfine field that is totally anisotropic. The hexagonal $HfFe_6Ge_6$-type structure of the $RMn_6Sn_{6-x}X_x$ family of compounds does not allow for such a cancellation, and both contributions to B_{hf} are present. In this case, the isotropic and anisotropic contributions cannot be distinguished unless a process is induced which would change only one of them. The spin reorientation process is a pure rotation of the magnetic structure relative to the crystal axes. Since the moments do not change either size or distance from the Sn site, nor does the magnetic symmetry change, the Fermi contact field cannot change under such a rotation. The isotropic contribution can therefore be isolated from the overall transferred hyperfine field at the Sn sites, and the anisotropic contribution can be determined uniquely.

A spin reorientation can be either temperature-induced or field-induced. At room temperature in $HfMn_6Sn_6$, the Mn moments are oriented in the *ab*-plane (the Mn sublattice has planar anisotropy). In $TbMn_6Sn_6$ at room temperature, the Tb and Mn moments lie along the *c*-axis, indicating the Tb sublattice to have uniaxial anisotropy which dominates over the planar anisotropy of the Mn sublattice. As the Tb anisotropy is more strongly temperature dependent, it decreases more rapidly on heating than that of the Mn. At T_{sr} = 330 K [4], the two sublattice anisotropies are equal, and above T_{sr} the planar Mn anisotropy dominates. The switch in dominant anisotropy causes the Tb and Mn moments to undergo a spontaneous reorientation from the *c*-axis to the *ab*-plane (on heating through T_{sr}). Alternatively, by applying a sufficient field perpendicular to the moment direction in a single crystal, we can induce a spin-flop to the direction along which the field is applied.

We have previously studied the anisotropic transferred hyperfine fields in $TbMn_6Sn_{6-x}Ga_x$ ($x = 0.2$, 0.4, 0.6 and 0.8) [5] through ^{119}Sn Mössbauer spectroscopy. This system undergoes a temperature-induced spin reorientation from the *ab*-plane to the *c*-axis on cooling through T_{sr}. The temperature at which the spin reorientation occurs was shown to decrease with increasing x. $TbMn_6Sn_{5.46}In_{0.54}$ single crystals undergo a spin-flop transition in quite modest fields around room temperature, and the single crystals can be grown as thin platelets suitable for Mössbauer studies. Here, we compare the spin-flop in $TbMn_6Sn_{5.46}In_{0.54}$ single crystals with the temperature-induced spin reorientation in $TbMn_6Sn_{6-x}Ga_x$ in order to demonstrate that the anisotropic contribution to the hyperfine field is independent of the driving force of the reorientation. We confirm that the Tb and Mn moments lie along *c* for $T < T_{sr}$ and use the electric field gradient to show that the spin reorientation in both systems involves a full rotation of the moments by 90°. The anisotropic contribution to B_{hf} is determined through this rotation.

2 Experimental methods

The $TbMn_6Sn_{6-x}Ga_x$ ($x = 0.2$, 0.4, 0.6 and 0.8) compounds were prepared by alloying stoichiometric amounts of ternary $TbMn_6Sn_6$ and $TbMn_6Ga_6$ compounds in an induction furnace [6]. The resulting ingots were sealed under argon in quartz tubes and annealed for 2 weeks at 973 K.

The $TbMn_6Sn_{5.46}In_{0.54}$ single crystals were synthesized using a flux method similar to that previously reported by Clatterbuck et al. [7]. A mixture of $TbMn_6Sn_6$ and indium and tin metal (99.9% pure) with the atomic ratio $TbMn_6Sn_{23}In_{37}$ was compacted into pellets and put into a silica tube with a quartz-wool stopper. The silica tube was sealed under argon (267 mbar) and heated to 1,273 K (at 50 K/h) for 24 h, after which it was cooled to 1,223 K (6 K/h). The sample was then re-heated to 1,263 K at the same rate and finally cooled slowly to 873 K in 65 h. The tube was quickly removed from the furnace, inverted and centrifuged manually using a David's sling device. The (Sn,In) flux settled to the bottom of the tube and the hexagonal crystal platelets (~1–2 mm in diameter and ~60 μm thick) remained on the quartz-wool stopper. Some of the platelets were ground and analyzed by x-ray diffraction Guinier patterns with Cu-K_α radiation. Both $TbMn_6Sn_{5.46}In_{0.54}$ and $TbMn_6Sn_{6-x}Ga_x$ are isotypic with the $HfFe_6Ge_6$ structure.

Mössbauer spectra were collected using a 10 mCi $^{119m}SnCaSnO_3$ source and the spectrometer was calibrated with α-Fe and a ^{57}Co source. Basic magnetic characterization was carried out on a commercial 9 T susceptometer/magnetometer. For the $TbMn_6Sn_{6-x}Ga_x$ samples, the temperature was varied from 12 to 300 K using a vibration-isolated closed-cycle refrigerator. The $TbMn_6Sn_{5.46}In_{0.54}$ sample was made of single crystal platelets with the c-axis perpendicular to the plane of the plates. The platelets were assembled in a mosaïc and subject to external magnetic fields from 0 to 1.53 T. The field was varied in a conventional electromagnet with the field perpendicular to both the γ-ray beam and the c-axis of the platelets. Spectra were fitted using a conventional non-linear least-squares minimization routine.

In order to study the reorientation process, a reference frame must first be identified. In a powder sample, the electric field gradient serves as a crystallographic reference frame. The high point symmetries of the three Sn sites in RMn_6Sn_6 ($6mm$ for Sn$-2e$ and $\bar{6}m2$ for Sn$-2c$ and Sn$-2d$) cause the crystallographic and electric field gradient axes to coincide, guaranteeing that the principal axis of the electric field gradient tensor V_{zz} lies along a, b or c. The local hexagonal point symmetries also impose axial symmetry ($\eta = 0$), such that the quadrupole splitting Δ is:

$$\Delta = \frac{eQV_{zz}}{4}(3\,cos^2\theta - 1) \tag{1}$$

where θ is the angle between V_{zz} and the hyperfine field at the Sn site due to the surrounding magnetic moments. For $\theta = 90^\circ$, $\Delta = -\frac{1}{4}eQV_{zz}$ and for $\theta = 0^\circ$, $\Delta = \frac{1}{2}eQV_{zz}$, so that a change in Δ by a factor of -2 corresponds to a change in moment direction by 90°.

The intensity of the second and fifth Mössbauer lines depends on the orientation of the magnetic moments with respect to the γ-ray (ϑ). Writing the intensity ratio as 3:R:1:1:R:3 for a six line Mössbauer pattern (for a $\frac{3}{2} \rightarrow \frac{1}{2}$ transition, such as in ^{119}Sn Mössbauer spectroscopy), ϑ can be determined:

$$R = \frac{4sin^2\vartheta}{1 + cos^2\vartheta} \tag{2}$$

For a single crystal, the moments point along a well defined direction with respect to the γ-direction. For example, the $TbMn_6Sn_{5.46}In_{0.54}$ single crystal platelets are grown

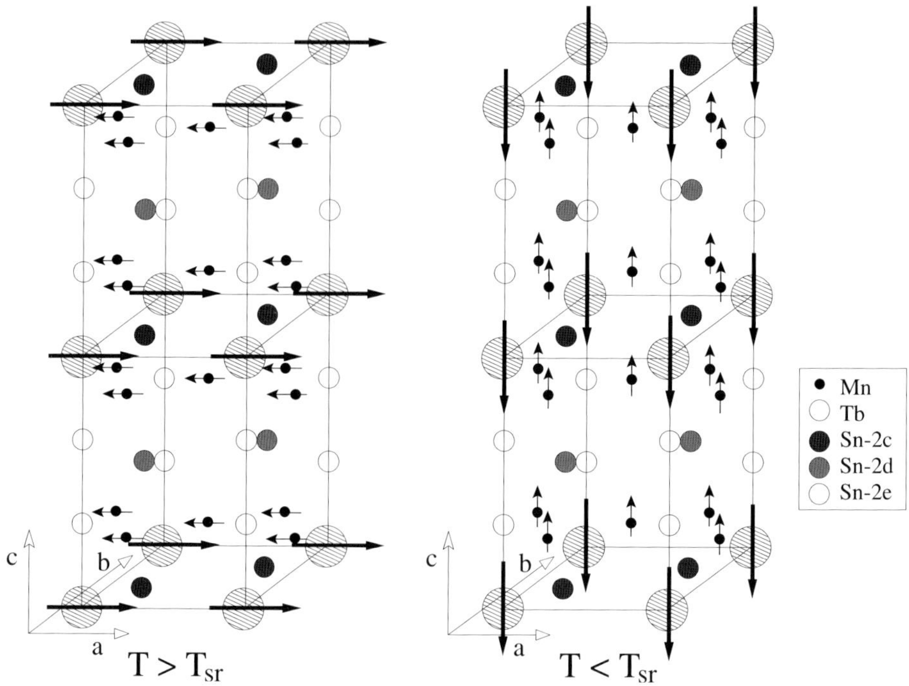

Figure 1 Crystal and magnetic structure of $TbMn_6Sn_6$ for temperatures above (*left*) and below (*right*) the reorientation temperature, T_{sr}.

with c perpendicular to the plane of the plates, parallel to γ. If the moments lie along the c-axis, then $\gamma \parallel c \parallel \mu$, giving $\vartheta = 0°$ and $R = 0$. If the moments are in the ab-plane, then $\gamma \parallel c \perp \mu$, giving $\vartheta = 90°$ and $R = 4$. For an oriented sample, R can be used to determine the moment direction. For a powdered sample, R averages to 2 and this information is lost.

3 Results

3.1 Temperature-induced spin reorientation in $TbMn_6Sn_{6-x}Ga_x$

The $TbMn_6Sn_6$ system has a $HfFe_6Ge_6$-type, $P6/mmm$ crystal structure, with lattice parameters $a = 5.53$ Å and $c = 9.023$ Å [8]. There are three crystallographic Sn sites in this structure: Sn$-2c$ (1/3, 2/3, 0), Sn$-2d$ (1/3, 2/3,1/2) and Sn$-2e$ (0, 0, 0.3356). The Sn$-2c$ and Sn$-2d$ sites both sit in the center of hexagonal Mn_6 prisms. Sn$-2c$ is in the plane of three Tb neighbours, while Sn$-2d$ has no rare earth neighbours [5, 8].[1] Sn$-2e$ lies in a hexagonal plane of Mn atoms, with one Tb neighbour above and

[1] In [5], the rare earth atom sits in the Tb$-1a$ site at (0,0,0). Sn$-2c$ has coordinates $(\frac{1}{3},\frac{2}{3},0)$ and so it is in the plane of three Tb atoms. In [8], the rare earth atom sits in the Tb$-1b$ site at $(0,0,\frac{1}{2})$. Thus Sn$-2d$ is labeled as the site in the plane of the rare earths instead of Sn$-2c$.

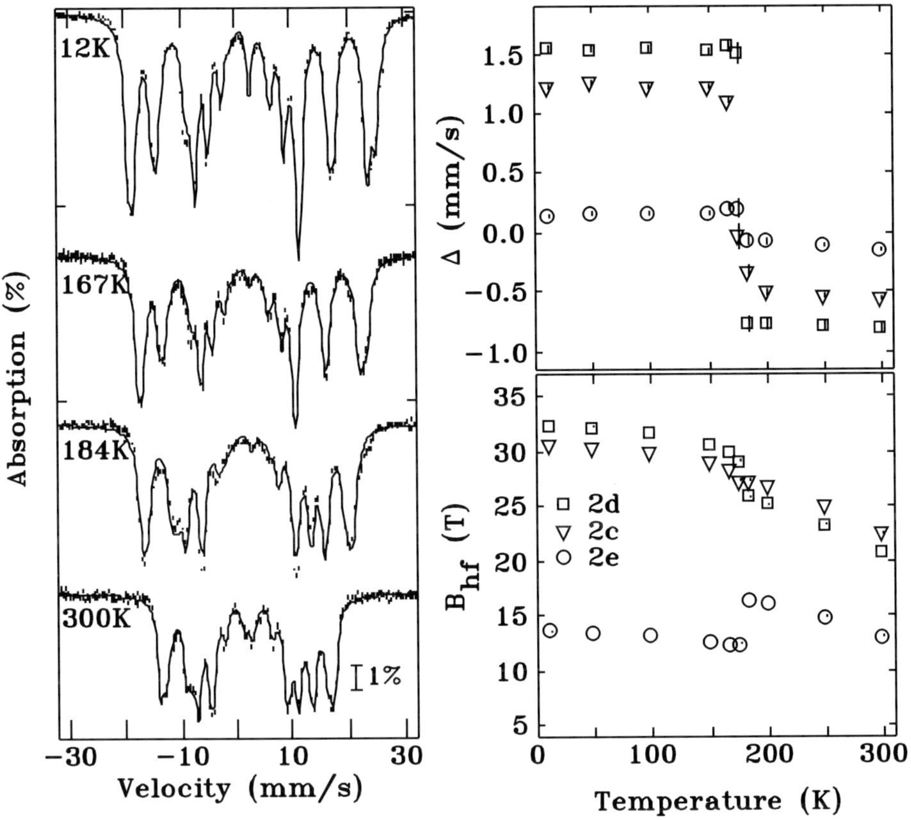

Figure 2 *Left*: ^{119}Sn Mössbauer spectra of $TbMn_6Sn_{5.4}Ga_{0.6}$ from 12 to 300 K. *Right*: quadrupole splitting (*top*) and hyperfine field (*bottom*) at the three Sn sites in $TbMn_6Sn_{5.4}Ga_{0.6}$, as functions of temperature. The spin reorientation temperature (T_{sr}) is 180(5) K.

one Sn neighbour below. Below T_N = 423 K, the magnetic structure of $TbMn_6Sn_6$ is characterized by alternating ferromagnetic Mn and Tb layers, ordered in the *ab*-plane and stacked along *c* [4] (Figure 1, left). The structure is ferrimagnetic due to strong antiferromagnetic coupling between the Mn and Tb layers. Cooling through T_{sr} = 330 K [4] causes the moments to spontaneously reorient onto the *c*-axis (Figure 1, right). The spin reorientation temperature (T_{sr}) is a strong function of Ga content (x), falling at 255 ± 18 K/Ga for increasing x [5].

Figure 2 (left) shows the Mössbauer spectra of $TbMn_6Sn_{5.4}Ga_{0.6}$ from 12 to 300 K. They consist of three subspectra (one for each Sn site) and a central Sn impurity. The assignment of each subspectrum to a Sn site cannot be based on the magnetic neighbours of the Sn sites. As shown in Figure 2 (bottom, right), two of the hyperfine fields (Sn−2*c* and Sn−2*d*) exchange sequence at the reorientation. For instance, if the largest field were to be associated with the site having the most magnetic neighbours, the site assignment would be different depending on whether the measurement were made above or below T_{sr}. Instead, the site assignment in [5] was based on the electrostatic environment (quadrupole splitting). This choice is not affected by their

Table I Mössbauer parameters at 167 and 200 K in $TbMn_6Sn_{5.4}Ga_{0.6}$

	$T < T_{sr}$ ($\mu \parallel c$) 167 K		$T > T_{sr}$ ($\mu \perp c$) 200 K	
Site	Δ (mm/s)	B_{hf} (T)	Δ (mm/s)	B_{hf} (T)
Sn−2*c*	1.11(3)	28.40(3)	−0.51(3)	26.73(3)
Sn−2*d*	1.56(4)	29.91(4)	−0.77(5)	25.18(5)
Sn−2*e*	0.21(3)	12.39(3)	−0.07(3)	16.20(2)

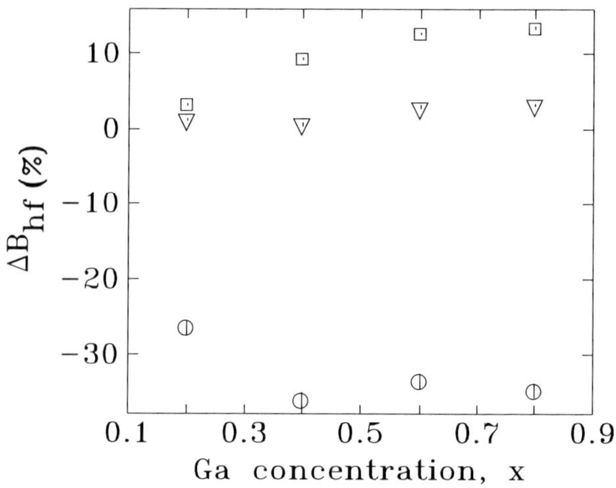

Figure 3 Fractional change in hyperfine field extrapolated to 0 K on two Brillouin curves, for $T < T_{sr}$ and $T > T_{sr}$.

orientation since the projection of the electric field gradient tensor changes by the same factor for all sites when the moments rotate. Table I lists the parameters for $T < T_{sr}$ and $T > T_{sr}$, where we see that the quadrupole splitting changes by a factor of −2 on cooling through T_{sr} = 180(5) K (Figure 2, top right). This corresponds to a change in moment direction of 90° (as discussed in Section 2), from the *ab*-plane for $T > T_{sr}$ to the *c*-axis for $T < T_{sr}$. The width of the transition is less than 5 K, so the reorientation is an abrupt, well-defined transition.

The temperature dependence of the hyperfine fields at the three Sn sites is shown in Figure 2 (bottom right). On cooling through T_{sr}, B_{hf} increases at Sn−2*d* and decreases at Sn−2*e*, and both are dramatic effects. In contrast, the hyperfine field at Sn−2*c* increases only slightly through the reorientation. The temperature dependence of B_{hf} was fitted to two Brillouin functions: one for $T < T_{sr}$ and another for $T > T_{sr}$. The two Brillouin curves were extrapolated to 0 K, and the difference δB_{hf} was normalized to the 0 K B_{hf} with $\mu \parallel c$. This gives the fractional change in hyperfine field, ΔB_{hf}, plotted in Figure 3 as a function of composition (Ga content, x). The fractional change in B_{hf} differs significantly for each site, and there is evidence for a slight increase in $|\Delta B_{hf}|$ for increasing Ga content.

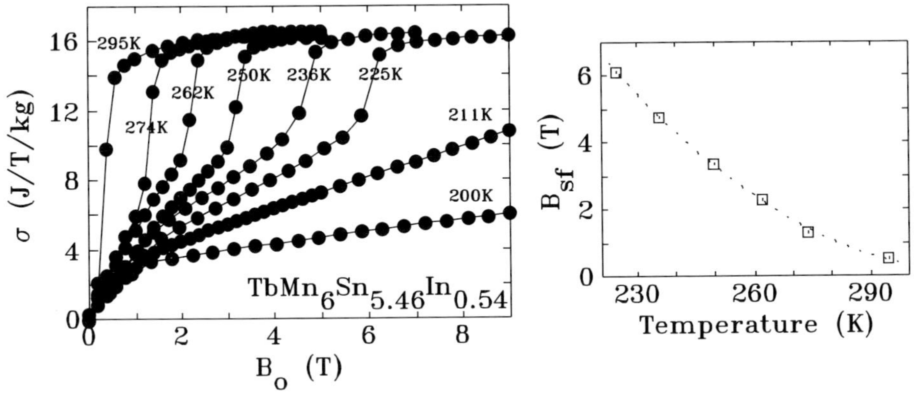

Figure 4 *Left*: Magnetization of $TbMn_6Sn_{5.46}In_{0.54}$ in an applied field range of 0 to 9 T and for $T < T_{sr}$. The applied field is perpendicular to the *c*-axis. *Right*: Temperature dependence of the spin-flop field in $TbMn_6Sn_{5.46}In_{0.54}$ for $T < T_{sr}$. The *dotted line* is a guide to the eye.

3.2 Field-induced spin-flop in $TbMn_6Sn_{5.46}In_{0.54}$

3.2.1 Magnetometry

The thermal variation of the magnetization of the single crystals measured above room temperature indicates the Néel point to be T_N= 397 K for $TbMn_6Sn_{5.46}In_{0.54}$. In addition, the compound displays a spin reorientation transition at T_{sr} = 317 K characterized by a change of the easy direction from the *ab*-plane at high temperature ($T > T_{sr}$) to the *c*-axis for $T < T_{sr}$.

Figure 4 (left) shows the magnetization of a $TbMn_6Sn_{5.46}In_{0.54}$ single crystal as a function of applied field, where B_o is perpendicular to *c*, and $T < T_{sr}$. In an applied field of ~0.5 T at room temperature, a spin-flop is induced to the direction along which B_o is applied (*ab*-plane). This confirms the moments are along the *c*-axis at room temperature. The spin-flop field increases with decreasing temperature (Figure 4, right). As the spin-flop field is expected to exceed the available 9 T at 220 K, the spin-flop is unattainable for 211 K and below.

3.2.2 Mössbauer spectroscopy

Figure 5 (left) shows the Mössbauer spectra for $TbMn_6Sn_{5.46}In_{0.54}$ in applied fields ranging from 0 to 1.53 T. In order to assign the subspectra to Sn sites as seen in Section 3.1, the sample was cooled to 12 K where the Mössbauer lines are more distinguishable. The site assignments at 12 K followed those of the $TbMn_6Sn_{6-x}Ga_x$ case [5] and sample parameters are shown in Table II. At 300 K, the quadrupole splittings were set to those determined at 12 K and the fields and isomer shifts were adjusted freely. The area constraint was 1:1:1.

For the single crystal platelets of $TbMn_6Sn_{5.46}In_{0.54}$, the moments are oriented along a unique direction with respect to the γ propagation direction. Through magnetometry it was shown that at 295 K and below, the moments are along the *c*-axis (parallel to γ). The intensity ratio *R* is plotted in Figure 5 (top right) as a function of applied field.

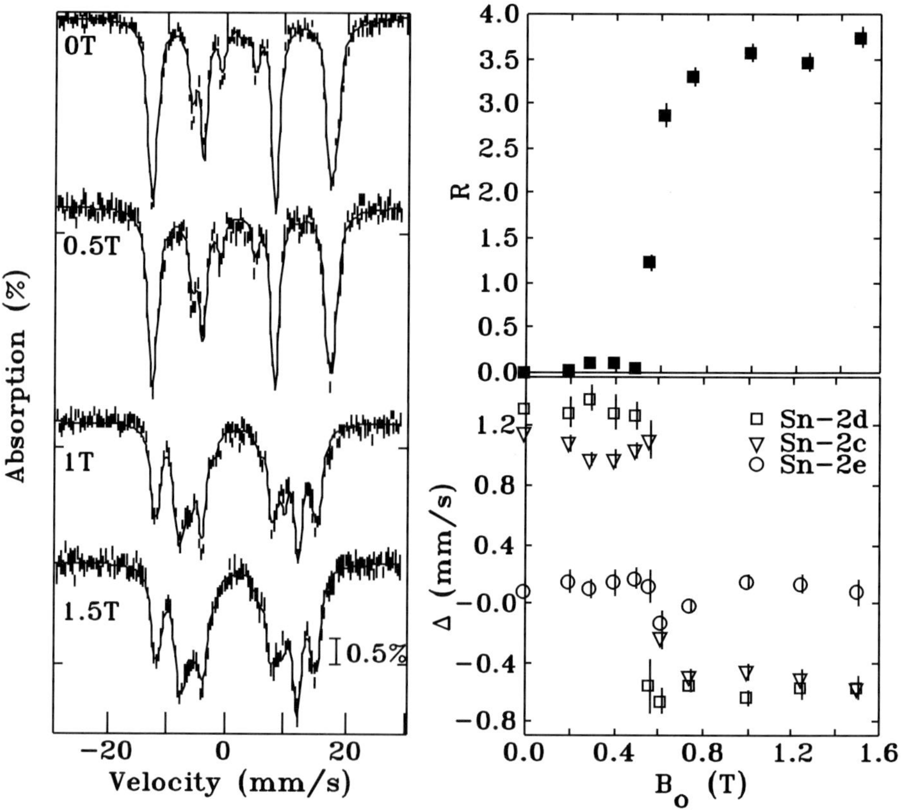

Figure 5 *Left*: ^{119}Sn Mössbauer spectra of $TbMn_6Sn_{5.4}Ga_{0.6}$ from 0 to 1.53 T. *Right*: R ratio (*top*) and quadrupole splitting (*bottom*) of the three Sn sites as functions of applied field. The spin-flop occurs at B_{sf} =0.57(3) T.

Table II Mössbauer parameters at 12 K in $TbMn_6Sn_{5.4}Ga_{0.6}$ and $TbMn_6Sn_{5.46}In_{0.54}$. The comparison of parameters between the two similar compounds guided the site assignment in $TbMn_6Sn_{5.46}In_{0.54}$

T = 12 K	$TbMn_6Sn_{5.4}Ga_{0.6}$ ($\mu \parallel c$)		$TbMn_6Sn_{5.46}In_{0.54}$($\mu \parallel c$)	
Site	Δ (mm/s)	B_{hf} (T)	Δ (mm/s)	B_{hf} (T)
Sn−2*c*	1.23(2)	30.57(2)	1.09(5)	26.26(2)
Sn−2*d*	1.55(3)	32.39(3)	1.33(9)	27.94(3)
Sn−2*e*	0.14(2)	13.64(2)	0.02(4)	12.23(2)

For $B_o < B_{sf}$, $R = 0$ confirms that the moments are parallel to the γ-rays ($\vartheta = 0°$) and thus to c. Once the spin-flop has occurred, an abrupt change in R from 0 to 3.7(1) is observed, indicating a change in moment direction. R is not 4 as expected for a full rotation into the ab-plane. This could either be due to an incomplete

Figure 6 Hyperfine fields at the three Sn sites in $TbMn_6Sn_{5.46}In_{0.54}$. The reorientation of the moments at 0.57(3) T on increasing applied fields causes an increase in B_{hf} at Sn−2*e*, and a decrease in B_{hf} at Sn−2*c* and Sn−2*d*.

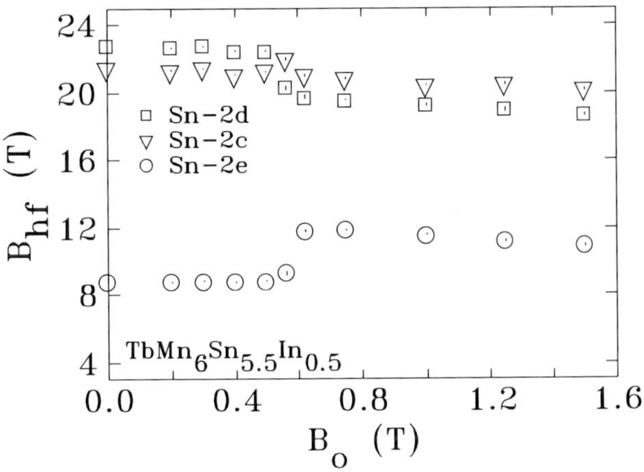

Table III In $TbMn_6Sn_{5.4}Ga_{0.6}$, the two $T < T_{sr}$ and $T > T_{sr}$ Brillouin curves are extrapolated to T = 0 K. The difference δB_{hf} is normalized to the 0 K B_{hf} with $\mu \parallel c$ to give the fractional change ΔB_{hf}. In $TbMn_6Sn_{5.46}In_{0.54}$, the two $B_o < B_{sr}$ and $B_o > B_{sr}$ lines of slope −1 are extrapolated to 0 T. The difference δB_{hf} is normalized to the 0 T B_{hf} with $\mu \parallel c$, giving ΔB_{hf}

	$TbMn_6Sn_{5.4}Ga_{0.6}$ T-induced		$TbMn_6Sn_{5.46}In_{0.54}$ B_o-induced	
Site	δB_{hf} (T)	ΔB_{hf} (%)	δB_{hf} (T)	ΔB_{hf} (%)
Sn−2*c*	0.6(2)	2.6(5)	0.02(2)	0.9(9)
Sn−2*d*	3.9(2)	12.7(4)	2.38(4)	10.4(2)
Sn−2*e*	−4.9(1)	−34(2)	−4.1(1)	−45(3)

rotation of the moments or to saturation effects. However, the factor of −2 change in quadrupole splitting (Figure 5, bottom right) associated with the spin-flop confirms that the rotation is complete. Therefore, the effect is likely due to the saturation of lines 2 and 5, caused by samples which are somewhat thick (the platelet thickness is ~60 μm).

Figure 6 shows the hyperfine fields at the three Sn sites as functions of applied field. For increasing B_o, the spin-flop is associated with an increase in B_{hf} at Sn−2*e*, and a decrease at Sn−2*c* and Sn−2*d*. These effects are identical to those observed in the thermally-driven spin reorientation in $TbMn_6Sn_{6-x}Ga_x$.The field dependence of B_{hf} was fitted to two lines of slope −1: one for $B_o < B_{sf}$ and another for $B_o > B_{sf}$. The lines were extrapolated to 0 T, and the difference δB_{hf} was normalized to the 0 T B_{hf} with $\mu \parallel c$ to give the fractional change in hyperfine field, ΔB_{hf}. In Table III, ΔB_{hf} is compared to that obtained in the thermally-driven reorientation in $TbMn_6Sn_{5.4}Ga_{0.6}$. The close agreement between the two measurements of ΔB_{hf} demonstrates that the anisotropic contribution can be obtained through both temperature-induced and field-induced spin reorientation and that it is independent of the driving force. The fractional change in hyperfine field due to the anisotropic contribution is large and negative in the case of Sn−2*e*, and smaller and positive at the Sn−2*c* and Sn−2*d* sites.

Table IV Vector sum over the six Mn neighbours ($\mathbf{A}_1$) for the three Sn sites in applied fields $B_o < B_{sf}$ and $B_o > B_{sf}$. The arrows in the third column indicate the magnitude A_1/A_p

	$\mathbf{A}_1/A_p$		
Site	$B_o < B_{sf}$ $\mu \parallel c$	$B_o > B_{sf}$ $\mu = (\frac{1}{\sqrt{2}}, \frac{1}{\sqrt{2}}, 0)\ \mu$	$\Delta A_1/A_p$
Sn$-2c$	$(0, 0, 3.97\mu)$	$(0.72\mu, 0.72\mu, 0) \rightarrow 1.01\ \mu$	$2.96\ \mu$
Sn$-2d$	$(0, 0, 4.02\mu)$	$(0.70\mu, 0.70\mu, 0) \rightarrow 0.99\mu$	$3.03\ \mu$
Sn$-2e$	$(0, 0, 0.46\mu)$	$(1.96\mu, 1.96\mu, 0) \rightarrow 2.77\mu$	$-2.31\ \mu$

Since the isotropic contribution to B_{hf} is independent of moment orientation, the changes in B_{hf} due to the spin-flop directly relate to the anisotropic contribution. The hyperfine field transferred to the Sn sites is given by [2, 3]:

$$\mathbf{B}_{hf} = \left(A_p \sum_{i=1}^{6} \mathbf{u}_i(\mu_i \cdot \mathbf{u}_i) - \frac{A_p}{3} \sum_{i=1}^{6} \mu_i \right) + A_s \sum_{i=1}^{6} \mu_i \tag{3}$$

where $\mathbf{u}_i$ is the unit vector connecting each Sn atom to a specific Mn atom with moment μ_i. A_p and A_s are the dipolar and contact fields due to a unit Mn moment. A_p is also referred to as the anisotropic constant. The last term in Eq. 3 is the isotropic contribution from the Mn shells around the Sn atoms, while the first two terms represent the anisotropic part of the transferred hyperfine field. For the spin-flop, we will only consider the first term ($\mathbf{A}_1$) of Eq. 3 involving the dot product, because the second term is independent of moment direction. The dot product $(\mu_i \cdot \mathbf{u}_i)$ is a maximum when the moment and Mn–Sn bond are parallel, and zero when they are perpendicular. From the crystal structure we know that the Mn–Sn bonds make angles of 35° (Sn$-2c$ and Sn$-2d$) and 73° (Sn$-2e$) with the c-axis. From the Mössbauer results discussed above (and from magnetometry and neutron scattering [9]), at $T < T_{sr}$ we know that the moments point along c for $B_o < B_{sf}$ and in the ab-plane for $B_o > B_{sf}$. As all the angles and orientations are known for any B_o, the anisotropic constant A_p can be determined.

Table IV lists the values for $\mathbf{A}_1$ at Sn$-2c$, Sn$-2d$ and Sn$-2e$ in applied fields above and below the spin-flop field. When the moments flop into the ab-plane from the c-axis, A_1 decreases at Sn$-2c$ and Sn$-2d$, while it increases at Sn$-2e$. Comparisons between $\Delta A_1/A_p$ and ΔB_{hf} show that despite their relative magnitudes being inconsistent, the signs of both quantities agree for all three Sn sites. The sign of $\Delta A_1/A_p$ for each Sn site mirrors the behaviour of the overall transferred hyperfine fields of Figure 6 and Table III, indicating that the anisotropic and isotropic contributions have the same sign. The Sn$-2e$ site was previously assigned by elimination in $TbMn_6Sn_{6-x}Ga_x$ [5]: Sn$-2c$ was assigned to the subspectrum which most reflected the values seen in $HfMn_6Sn_6$ [1], the Sn$-2d$ site assignment was argued based on similarity to Sn$-2c$, and the remaining subspectrum was assigned Sn$-2e$. Here we see that Sn$-2e$ is unique because it has a negative $\Delta A_1/A_p$ and ΔB_{hf}. This distinction allows the Sn$-2e$ site to be directly identified and confirms that the site assignment in $TbMn_6Sn_{6-x}Ga_x$ was correct.

In the tetragonal $MnSn_2$ system ($I4/mcm$, $a = 6.659$ Å, $c = 5.447$ Å, with Mn$-4a$ at (0, 0, 0.25) and Sn$-8h$ at (0.1623,0.6623, 0)), there is no spin reorientation or

Table V Calculated anisotropic contributions to B_{hf} at the three Sn sites in $TbMn_6Sn_{5.46}In_{0.54}$. The anisotropic field B^A_{hf} is equal to δB_{hf} from Table III

Site	$\Delta A_1/A_p$ (μ)	B^A_{hf} (T)	A_p (T/μ_B)
Sn–2*c*	2.96	0.02(2)	0.003(2)
Sn–2*d*	3.03	2.38(4)	0.39(2)
Sn–2*e*	−2.31	−4.1(1)	0.89(5)

spin-flop, and ΔA_1 is simply A_1. The size of A_1/A_p is 0.96 μ and the magnitude of the Mn moment is $\mu = 2.33\ \mu_B$ [2]. The antiferromagnetic environment of the Sn_{00} site guarantees that the transferred hyperfine field at Sn_{00} is purely anisotropic. The anisotropic transferred hyperfine field B^A_{hf} relates to the anisotropic constant A_p through:

$$A_p = \frac{B^A_{hf}}{\Delta A_1} \tag{4}$$

At Sn_{00}, $B^A_{hf} = 4.9(2)$ T, giving the anisotropic constant $A_p = 2.19(1)\ T/\mu_B$ [2]. For the spin-flop in $TbMn_6Sn_{5.46}In_{0.54}$, the anisotropic field B^A_{hf} is taken as the difference between the $\mu \parallel c$ and $\mu \perp c$ hyperfine fields extrapolated to $B_o = 0$ T (δB_{hf} from Table III). From room temperature neutron diffraction studies of $TbMn_6Sn_6$, we know that the Mn moment is 1.99(6) μ_B [9]. The anisotropic constant A_p in $TbMn_6Sn_{5.46}In_{0.54}$ can therefore be determined for all three Sn sites (Table V).

The anisotropic constants at the Sn sites in $TbMn_6Sn_{5.46}In_{0.54}$ are significantly smaller than A_p measured at the Sn_{00} site in $MnSn_2$ (which is the only site allowing for the cancellation of the isotropic contribution). As the anisotropic contribution is related to the localization of electrons on the Mn–Sn bond, and therefore to covalency effects [2], the smaller anisotropic constants seen here could suggest that the Mn–Sn bonds in $TbMn_6Sn_{5.46}In_{0.54}$ are less covalent. Parasitic anisotropic effects arising from Tb–Sn bonding could also reduce the Mn–Sn anisotropic contribution in this system, leading to smaller anisotropic constants.

4 Conclusions

The Mössbauer study of the temperature-induced spin reorientation and the field-induced spin-flop (in $TbMn_6Sn_{6-x}Ga_x$ and $TbMn_6Sn_{5.46}In_{0.54}$ respectively) has demonstrated that the anisotropic contribution to the transferred hyperfine field is independent of the force which drives the reorientation. We have used magnetometry and Mössbauer spectroscopy to show that the moments are along the *c*-axis for $T < T_{sr}$. The change in anisotropic field at the reorientation is seen to be largest at the Sn–2*e* site, and the identification of this site is validated. The anisotropic contribution has been shown to be substantial, though less than in the $MnSn_2$ system, a possible consequence of weaker covalent bonding or contributions from Tb–Sn bonds. Future investigations of non-magnetic rare earth compounds will be performed with the intent of evaluating the possibility of a parasitic anisotropic effect from R–Sn (R = magnetic rare earth) bonding. The replacement of magnetic

Tb with a non-magnetic rare earth atom also allows for further studies of covalency effects in the RMn_6Sn_6 system.

References

1. Mazed, T., Tobola, J., Venturini, G., Malaman, B.: Phys. Rev., B **65**, 104406 (2002)
2. Le Caër, G., Malaman, B., Venturini, G., Kim, I.B.: Phys. Rev., B **26**, 5085 (1982)
3. Le Caër, G., Malaman, B., Venturini, G., Fruchart, D., Roques, B.: J. Phys. F. Met. Phys. **15**, 1813 (1985)
4. Venturini, G., Chafik El Idrissi, B., Malaman, B.: J. Magn. Magn. Mater. **94**, 35 (1991)
5. Perry, L.K., Ryan, D.H., Venturini, G., Cadogan, J.: J. Appl. Phys. **99**, (2006)
6. Lefèvre, C., Venturini, G., Malaman, B.: J. Alloys Compd. **346**, 84 (2002)
7. Clatterbuck, D.M., Gschneidner, K.A. Jr.: J. Magn. Magn. Mater. **207**, 78 (1999)
8. Chafik El Idrissi, B., Venturini, G., Malaman, B.: Mater. Res. Bull. **26**, 431 (1991)
9. Chafik El Idrissi, B., Venturini, G., Malaman, B.: J. Less-common Met. **175**, 143 (1991)

Hyperfine Interact (2006) 170:117–125
DOI 10.1007/s10751-006-9465-4

Use of Mössbauer spectroscopy to determine the effect of salinity on the speciation of triorganotins in Anacostia River sediments

George Eng · Xueqing Song · Leopold May

Published online: 9 January 2007

Abstract The speciation of several tributyltin and triphenyltin compounds under varying salinity conditions (0, 20, 40 and 60%) was studied by Mössbauer spectroscopy in both anaerobic and aerobic Anacostia River sediments. The Mössbauer spectral parameters of the spiked sediments indicated that changes in the salinity did not affect the speciation of the tin compounds in either aerobic or anaerobic sediments.

Key words antifoulant · Mössbauer spectroscopy · speciation · Anacostia River · sediments · tributyltins · triorganotins · triphenyltins

1 Introduction

An organotin compound is one that contains at least one tin–carbon bond. In recent years, there has been an increased usage of organotins due to their numerous biocidal properties. In fact, there are more commercial uses of organotins than of any other organometallic system [1]. Tributyltins (TBTs) and triphenyltins (TPTs) are commonly used as the active ingredient in marine paints [2]. Although the use of these triorganotins as antifoulants prevents biofouling, it permits the direct introduction of these toxicants into various aquatic

G. Eng · X. Song
Department of Chemistry and Physics and Agricultural Experimental Station,
University of the District of Columbia,
Washington, DC 20008, USA

G. Eng
Agricultural Experimental Station, University of the District of Columbia,
Washington, DC 20008, USA

L. May (✉)
Department of Chemistry, The Catholic University of America,
Washington, DC 20064, USA
e-mail: may@cua.edu

Table 1 Mössbauer spectral parameters of TBTCl in the Anacostia River anaerobic and aerobic sediments with varying salinities[a]

Site	Salinity	Δ		δ		ρ	
		Anaerobic	Aerobic	Anaerobic	Aerobic	Anaerobic	Aerobic
1	0	3.32	3.22	1.44	1.45	2.31	2.22
	20	3.19	3.34	1.41	1.46	2.26	2.29
	40	3.43	3.36	1.44	1.50	2.39	2.24
	60	3.38	3.38	1.44	1.50	2.35	2.25
2	0	3.31	3.46	1.45	1.48	2.28	2.34
	20	3.42	3.40	1.47	1.44	2.32	2.36
	40	3.40	3.49	1.44	1.48	2.35	2.36
	60	3.34	3.46	1.41	1.47	2.37	2.35
3	0	3.30	3.49	1.38	1.47	2.38	2.37
	20	3.34	3.55	1.40	1.52	2.40	2.34
	40	3.36	3.51	1.44	1.49	2.34	2.36
	60	3.28	3.51	1.45	1.51	2.26	2.32
4	0	3.38	3.40	1.43	1.44	2.36	2.36
	20	3.52	3.45	1.49	1.45	2.36	2.38
	40	3.48	3.50	1.49	1.48	2.34	2.36
	60	3.53	3.45	1.51	1.49	2.34	2.32
5	0	3.29	3.41	1.46	1.45	2.25	2.35
	20	3.40	3.42	1.44	1.41	2.37	2.43
	40	3.52	3.44	1.49	1.44	2.37	2.39
	60	3.45	3.45	1.46	1.47	2.36	2.35
Average		3.38(7)	3.44(7)	1.45(3)	1.47(2)	2.34(4)	2.34(5)
Pure		3.43		1.56		2.20	

[a] All values relative to $BaSnO_3$ at RT in mm/s. The numbers in the parentheses are the errors in the last figures.

ecosystem [3]. It has been shown that TBTs have adverse effects on non-targeted species such as oysters [4]. The use of triorganotin compounds in the United States has been restricted by the Organotin Act, which prohibits the use of organotin based paints on vessels smaller than 25 m. However, vessels larger than 25 m may still use marine paints containing organotins [5]. Even with the restricted use of these antifouling marine paints, these compounds have already entered various water systems during their previously unrestricted use. Thus, it is important to determine the fate of these triorganotins in the marine hydrosphere.

There are numerous analytical procedures in the literature for the determination of organotin compounds. Two reviews [6, 7] as well as one oriented towards environmental issues have indicated that the method most employed for the determination of organotin species involves extraction and derivatization of the organotin species followed by species detection [8]. However, strong interactions between triorganotins and sediments can bias the results [7]. Furthermore, the accuracy for butyl- and phenyltin determination is hampered by the lack of certified reference materials [7]. It would also be more advantageous to examine the organotin species directly in the sediment than to study their derivatized analogs because metals [9], organics or sulfides contained in the sediment [10, 11] can interfere with the derivatization of the organotin species. Mössbauer spectroscopy permits direct measurements of the organotin species in the sediment thus eliminating

Table 2 Mössbauer spectral parameters of TBTO in Anacostia River anaerobic and aerobic sediments with varying salinities[a]

Site	Salinity	Δ		δ		ρ	
		Anaerobic	Aerobic	Anaerobic	Aerobic	Anaerobic	Aerobic
1	0	3.27	3.22	1.41	1.45	2.32	2.22
	20	3.25	3.34	1.41	1.46	2.30	2.29
	40	3.27	3.36	1.41	1.50	2.31	2.24
	60	3.30	3.38	1.42	1.50	2.32	2.25
2	0	3.33	3.46	1.45	1.48	2.30	2.34
	20	3.35	3.40	1.42	1.44	2.37	2.36
	40	3.38	3.49	1.48	1.48	2.29	2.36
	60	3.28	3.46	1.44	1.47	2.27	2.35
3	0	3.20	3.49	1.39	1.47	2.30	2.37
	20	3.22	3.55	1.42	1.52	2.27	2.34
	40	3.28	3.51	1.43	1.49	2.29	2.36
	60	3.25	3.51	1.44	1.51	2.25	2.32
4	0	3.29	3.40	1.42	1.44	2.33	2.36
	20	3.24	3.45	1.39	1.45	2.33	2.38
	40	3.31	3.50	1.42	1.48	2.33	2.36
	60	3.36	3.45	1.44	1.49	2.33	2.32
5	0	3.36	2.88	1.45	1.34	2.32	2.15
	20	3.26	3.09	1.43	1.41	2.28	2.19
	40	3.13	3.09	1.40	1.41	2.24	2.19
	60	3.39	3.15	1.35	1.42	2.51	2.22
Average		3.29(4)	3.26(19)	1.42(2)	1.46(4)	2.31(6)	2.30(7)
Pure		1.89		1.24		1.53	

[a] All values relative to $BaSnO_3$ at RT in mm/s. The numbers in the parentheses are the errors in the last figures.

problems associated with extraction. In the hopes of getting a more complete picture of the speciation of triorganotins in the Anacostia River, a salinity study was undertaken. The results of the Mössbauer spectral studies from the spiking of anaerobic and aerobic sediments from the Anacostia River with triorganotin compounds that are commonly incorporated into marine paints as a function of salinity are reported here.

2 Experimental

Tributyltin chloride (TBTCl) and *bis*-tributyltin oxide (TBTO) were obtained from M & T Chemicals, Inc., Rahway, NJ, USA. Tributyltin acetate (TBTOAc) and triphenyltin hydroxide (TPTOH) were purchased from Gelest, Inc., Tullytown, PA, USA and triphenyltin chloride (TPTCl) was obtained from Aldrich Chemical Co., Inc., Milwaukee, WI, USA. Triphenyltin acetate (TPTOAc) was obtained from Alfa Products, Danvers, MA, USA. All the compounds contained the normal abundance of ^{119}Sn and were used as received without further purification to spike the sediment samples.

Sediment samples were obtained as grab samples from the Anacostia River in the DC metropolitan area. The samples were kept frozen until they were ready to be spiked. Aerobic sediments were prepared by allowing the anaerobic sediments to dry in air. The

Table 3 Mössbauer spectral parameters of TBTOAc in Anacostia River anaerobic and aerobic sediments with varying salinities[a]

Site	Salinity	Δ		δ		ρ	
		Anaerobic	Aerobic	Anaerobic	Aerobic	Anaerobic	Aerobic
1	0	3.28	3.22	1.41	1.44	2.33	2.24
	20	3.45	3.13	1.50	1.40	2.30	2.24
	40	3.06	3.27	1.46	1.45	2.10	2.26
	60	3.40	3.23	1.51	1.44	2.25	2.24
2	0	3.28	3.14	1.44	1.41	2.27	2.23
	20	3.39	3.17	1.47	1.40	2.31	2.26
	40	3.37	3.21	1.46	1.44	2.31	2.23
	60	3.33	3.33	1.42	1.45	2.35	2.30
3	0	3.28	3.36	1.37	1.41	2.39	2.38
	20	3.10	3.33	1.37	1.45	2.26	2.30
	40	3.22	3.36	1.38	1.43	2.33	2.35
	60	3.22	3.34	1.38	1.46	2.33	2.29
4	0	3.31	3.30	1.43	1.41	2.31	2.34
	20	3.36	3.25	1.40	1.41	2.39	2.30
	40	3.43	3.24	1.48	1.39	2.31	2.33
	60	3.41	3.23	1.47	1.44	2.32	2.24
5	0	3.33	3.29	1.45	1.41	2.29	2.33
	20	3.39	3.47	1.45	1.47	2.33	2.36
	40	3.33	3.29	1.42	1.44	2.34	2.28
	60	3.30	3.39	1.43	1.47	2.30	2.31
Averages		3.31(7)	3.28(7)	1.44(4)	1.43(1)	2.31(6)	2.29(5)
Pure		3.51		1.46		2.40	

[a] All values relative to $BaSnO_3$ at RT in mm/s. The numbers in the parentheses are the errors in the last figures.

color of the sediments changed from black/greenish to black to brown. The locations of the various sites have been previously reported [12].

Salinity of the anaerobic or aerobic sediments was adjusted with artificial seawater so that the resulting salinities of the samples were 20, 40, and 60% of that of synthetic seawater, which was taken as 100%. Anaerobic sediments were thawed in water to prevent oxidation. The following procedure was used in all experiments. Five grams of sediment were spiked with 0.1 g of the appropriate triorganotin compound. The mixture was then covered with 100 ml of water of the appropriate salinity. The mixture was shaken mechanically in a closed vessel in the dark for 2 weeks at room temperature and remained in the dark at room temperature for an additional week. The sediment samples were then removed by gravity filtration and kept frozen until the Mössbauer spectra were recorded.

The Mössbauer spectra were recorded using a Model MS-900 (Ranger Scientific Co., Burleson, TX, USA) spectrometer in the acceleration mode with moving source geometry. A 10 mCi $Ca^{119m}SnO_3$ source was used, and counts of 30,000 or more were accumulated for each spectrum. The spectra were measured at 80 K using a liquid–nitrogen cryostat (CRYO Industries of America, Inc., Salem, NH, USA). The velocity was calibrated at ambient temperature using a composition of $BaSnO_3$ and tin foil (splitting 2.52 mm s^{-1}). The resultant spectra were analyzed by a least-square fit to Lorentzian shaped lines.

Table 4 Mössbauer spectral parameters of TPTCl in Anacostia River anaerobic and aerobic sediments with varying salinities[a]

Site	Salinity	Δ		δ		ρ	
		Anaerobic	Aerobic	Anaerobic	Aerobic	Anaerobic	Aerobic
1	0	2.71	2.79	1.17	1.16	2.32	2.40
	20	2.84	2.82	1.20	1.18	2.37	2.39
	40	2.75	2.79	1.18	1.19	2.33	2.34
	60	2.68	2.43	1.14	1.20	2.35	2.03
2	0	2.87	2.45	1.21	1.13	2.38	2.20
	20	2.71	2.50	1.24	1.21	2.19	2.07
	40	2.70	2.45	1.24	1.20	2.17	2.04
	60	2.74	2.71	1.21	1.11	2.28	2.44
3	0	2.88	2.77	1.20	1.16	2.40	2.39
	20	2.83	2.66	1.19	1.20	2.37	2.22
	40	2.83	2.71	1.19	1.21	2.38	2.24
	60	2.77	2.71	1.19	1.24	2.33	2.19
4	0	2.86	2.61	1.19	1.16	2.41	2.31
	20	2.74	2.76	1.20	1.18	2.29	2.38
	40	2.79	2.67	1.19	1.20	2.34	2.23
	60	2.75	2.61	1.21	1.17	2.27	2.23
5	0	2.90	2.42	1.23	1.06	2.36	2.28
	20	2.78	2.61	1.08	1.12	2.56	2.33
	40	2.70	2.58	1.15	1.11	2.35	2.32
	60	2.86	2.71	1.20	1.18	2.38	2.29
Averages		2.79(4)	2.64(8)	1.19(2)	1.17(3)	2.34(8)	2.27(13)
Pure		2.53		1.31		1.93	

[a] All values relative to $BaSnO_3$ at RT in mm/s. The numbers in the parentheses are the errors in the last figures.

3 Results and discussion

The Mössbauer spectral parameters, quadrupole splitting (Δ) and isomer shift (δ), for the sediments spiked with TBT compounds at the different salinities are listed in Tables 1, 2 and 3 and with those spiked with TPT compounds are given in Tables 4, 5 and 6. Typical spectra for the spiked aerobic and anaerobic sediments are shown in Fig. 1.

The average Δ value calculated for the anaerobic and aerobic sediments spiked with TBTCl is equal to the Δ value of the pure compound; however, the δ value for the sediments is slightly smaller than the δ value of the pure TBTCl (Table 1). The spectral parameters of the spiked sediments are within experimental error of each other suggesting that in both types of sediments they are being converted to the same species, most likely the hydrated tributyltin cation. Similar results have been reported for the speciation of TBTs in Chesapeake Bay sediments [13]. In addition, the parameters at the different sites for the different salinities are similar suggesting that salinity did not play a major role in the speciation.

The Mössbauer parameters for pure TBTO are smaller than the corresponding parameters found in the individual spectra of the spiked sediments as well as those for the average values (Table 2). This would clearly indicate that some type of speciation has

Table 5 Mössbauer spectral parameters of TPTOH in Anacostia River anaerobic and aerobic sediments with varying salinities[a]

Site	Salinity	Δ		δ		ρ	
		Anaerobic	Aerobic	Anaerobic	Aerobic	Anaerobic	Aerobic
1	0	2.82	2.82	1.18	1.18	2.40	2.39
	20	2.81	2.79	1.17	1.19	2.40	2.34
	40	2.77	2.80	1.18	1.17	2.35	2.39
	60	2.79	2.85	1.18	1.17	2.37	2.44
2	0	2.85	2.87	1.20	1.19	2.38	2.41
	20	2.85	2.85	1.19	1.20	2.39	2.38
	40	2.81	2.85	1.16	1.19	2.43	2.39
	60	2.88	2.85	1.23	1.20	2.35	2.38
3	0	2.80	2.85	1.16	1.15	2.41	2.48
	20	2.84	2.83	1.18	1.14	2.41	2.48
	40	2.86	2.88	1.18	1.15	2.41	2.50
	60	2.80	2.86	1.21	1.16	2.32	2.47
4	0	2.82	2.86	1.16	1.18	2.43	2.42
	20	2.84	2.85	1.17	1.20	2.44	2.38
	40	2.82	2.79	1.17	1.16	2.42	2.41
	60	2.81	2.74	1.16	1.21	2.42	2.26
5	0	2.73	2.84	1.16	1.15	2.36	2.47
	20	2.81	2.84	1.16	1.17	2.41	2.43
	40	2.78	2.85	1.14	1.19	2.44	2.39
	60	2.84	2.86	1.21	1.16	2.35	2.47
Averages		2.81(3)	2.84(2)	1.18(1)	1.18(2)	2.39(3)	2.41(6)
Pure		2.84		1.18		2.41	

[a] All values relative to $BaSnO_3$ at RT in mm/s. The numbers in the parentheses are the errors in the last figures.

occurred within the two types of sediments. Based on the Mössbauer parameters, it is speculated that the TBTO hydrolyzed in water and formed TBTOH, which can then interact with the sediment. With some exceptions, both the Δ and δ values are the same at different salinities suggesting again, that salinity did not affect the speciation.

In sediments spiked with TBTOAc, the Δ value is higher in the pure compound than in the spiked sediments, but the δ-values are within experimental error suggesting some type of speciation has again occurred. It is postulated that TBTOAc dissociates to the hydrated TBT^+ cation and then interact with the sediments. In addition, Table 4 shows that the Mössbauer parameters do not change as a function of salinity.

In summary, the TBT data reveal that with the exception of the aerobic sediment spiked with TBTCl, the spectral data for all the other tributyltin compounds in both types of sediments are the similar within experimental error. This would suggest that the compounds are converted to a common species, most likely the hydrated tributyltin cation. An equation that would fit this hypothesis is: $TBTX + nH_2O = TBT(H_2O)_n^+ + X^-$, where $X{=}Cl^-$, OH^-, OAc^-. An earlier study reported similar findings. [13] The hydrated species can then interact with the sediment according to the following equation: $TBT(H_2O)_n^+ + \text{sediment} =$ $TBT(H_2O)_n^+ - \text{sediment}$ [13]. In general, the data for the TBT compounds indicated that anaerobic and aerobic spiked sediments from the different sites under the various salinity

Table 6 Mössbauer spectral parameters of TPTOAc in Anacostia River anaerobic and aerobic sediments with varying salinities[a]

Site	Salinilty	Δ		δ		ρ	
		Anaerobic	Aerobic	Anaerobic	Aerobic	Anaerobic	Aerobic
I	0	1.95	2.82	0.63	1.18	3.08	2.39
	20	1.90	2.79	0.62	1.19	3.05	2.34
	40	2.05	2.80	0.64	1.17	3.20	2.39
	60	2.02	2.85	0.64	1.17	3.14	2.43
II	0	2.04	2.87	0.69	1.19	2.96	2.41
	20	1.99	2.85	0.64	1.20	3.14	2.38
	40	1.95	2.85	0.63	1.19	3.12	2.39
	60	1.99	2.85	0.62	1.20	3.24	2.38
III	0	2.01	2.85	0.62	1.15	3.24	2.48
	20	1.99	2.83	0.65	1.14	3.07	2.48
	40	2.07	2.88	0.65	1.15	3.17	2.50
	60	2.05	2.86	0.70	1.16	2.95	2.47
IV	0	1.63	2.84	0.45	1.18	3.62	2.41
	20	1.76	2.85	0.64	1.20	2.75	2.38
	40	1.67	2.79	0.78	1.16	2.14	2.41
	60	1.46	2.74	0.53	1.21	2.75	2.26
V	0	1.95	2.80	0.58	1.19	3.36	2.35
	20	1.74	2.84	0.68	1.18	2.56	2.41
	40	1.60	2.82	0.64	1.18	2.50	2.39
	60	2.09	2.85	0.84	1.17	2.49	2.44
Averages		2.06(16)	2.83(2)	0.62(3)	1.18(2)	2.98 (35)	2.41(6)
Pure		3.35		1.28		2.61	

[a] All values relative to $BaSnO_3$ at RT in mm/s. The numbers in the parentheses are the errors in the last figures.

conditions gave similar Δ and δ values. This would suggest that salinity did not play a major role in the speciation of the compounds. This result is similar to a study involving Chesapeake Bay sediments [13].

Rho values (Δ/δ)have been used to deduce the coordination number of tin atoms in various organotin compounds. Values less than 2.1 have been identified with compounds in which the tin atom is less than four-coordinated, while five-and higher coordination has been assigned to values greater than 2.1 [14]. As can be seen from Tables 1, 2 and 3, the rho values range from 2.22 to 2.43 indicating that the coordination for the tin atom in all the spiked sediments (anaerobic and aerobic) is greater than 4. This would indicate that there is some type of binding between the tributyltins and the anaerobic and aerobic sediments supporting the above conclusions.

As can been seen in Table 5, the spectral parameters for TPTOH in the spiked sediments is the same as the parent compounds. This would suggest that the TPTOH did not change in the spiked sediments. However, there are significant differences in the averaged Δ and δ values in the sediments spiked with TPTCl and TPTOAc and the respective parent compounds. It is further noted that the anaerobic and aerobic sediments spiked TPTCl as well as the aerobic sediments spiked with TPTOAc have spectral parameters similar to the values observed for TPTOH. This would suggest that the compounds in these environments

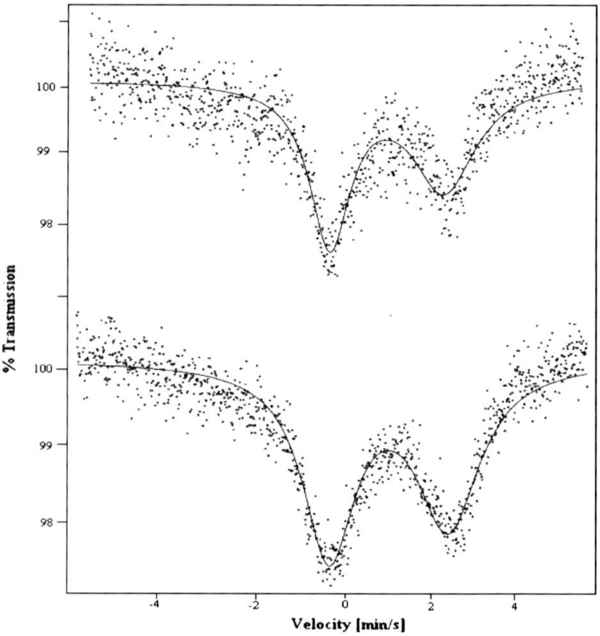

Fig. 1 Mössbauer spectra of triphenyltin chloride (*TPTCl*) in anaerobic (*top*) and aerobic (*bottom*) spiked sediments from site 4 at salinity of 40%.

form the same species in the various types of sediments. Similar to the tributyltin case, the species would be the hydrated triphenyltin cation, and it is this species that would interact with the sediment. Similar results have been previously report [13].

The one inconsistent result in the present study is in the anaerobic sediments spiked with TPTOAc. The spectral parameters are substantially different from all the other results. A comparison of the spectral parameters to those in the literature reveal that the parameters in the spiked TPTOAc anaerobic sediments (Δ=2.0 mm s^{-1}; δ=0.6 mm s^{-1}) are similar to those found for diphenyltin dichloride (Δ=2.0 mm s^{-1}; δ=1.0 mm s^{-1}). Thus, it is hypothesized that a dephenylation of the TPTOAc has occurred and that a diphenyltin species has formed. Various microorganisms found in anaerobic sediments can achieve the dephenylation. Degradation of organotin compounds due to various bacteria or macroalgae in sediments has been reported in the literature [15].

Rho values for the anaerobic and aerobic sediments ranged from 2.03 to 2.56 for samples spiked with TPTCl, 2.32 to 2.50 for samples spiked with TPTOH, and 2.14 to 3.62 for samples spiked with TPTOAc. This would indicate, similar to the TBT results, that in all cases the coordination number of the tin atom of the compounds in the spiked sediments is greater than four, indicating that some type of binding has occurred between all these triphenyltin compounds and the sediments.

A comparison of the spectral parameter between the TPT and TBT compounds indicate that the parameters for the TBT spiked sediments are larger than those found for the TPT compounds. The smaller Δ values observed for the TPT compounds would indicate that the environment around the tin atom in these compounds in the sediments is more symmetrical. Smaller δ values observed for the TPT compounds would indicate that the interactions of the hydrated TPT$^+$ cation with the sediments are stronger than in case for the TPT compounds since δ values are known to vary with the bonding of the tin atom. A comparison of the phenyl group to the butyl group can explain this observation. The steric effect of A more flexible n-butyl group would prevent the hydrated TBT^+ cation from having strong contacts with the sediments as have been previously reported [13].

Similar to the TBT results, there was not significant differences in the Mössbauer parameters at the same site under different salinity conditions. This indicates that salinity does not appear to be a major factor influencing the speciation of the TPT compounds. The results from this study are similar to those reported earlier in the Chesapeake Bay study [16].

In addition, Anacostia River sediments are heterogeneous mixtures of clays, sand, and other materials whose composition varies along the river floor [17]. Hence, any variations in the Mössbauer parameters at the different sites may reflect the dissimilarity in the components of the sediments as well as the manner in which these components interact with the different triorganotins. Thus, a more complete understanding of these variations awaits the total analyses of the components/structures of the sediments in the Anacostia River.

References

1. Blunden, S.J., Cusack, P.A., Hill, R.: The Industrial Uses of Tin Chemicals. Burlington House, London, p. 1 (1985)
2. Davis, A.G., Smith, P.J.: In: Wilkinson, G., Stone, F.G.A., Abel, E.W. (eds.) Comprehensive Organometallic Chemistry, vol. 2, p. 519. Pergamon, New York, (1982)
3. Blunden, S.J., Hobbs, L.A., Smith, P.J.: In: Bowen, H.J.M. (ed.) Environmental Chemistry, p. 49. The Royal Society of Chemistry, London (1984)
4. Hall, K.W. Jr, Pinkney, A.E.: CRC Crit. Rev. Toxicol. **14**, 159 (1985)
5. Fed. Reg. **53**, 39022 (1988)
6. Leroy, M.J.F., Quevauviller, P., Donard, O.F.X., Astruc, M.: Pure Appl. Chem. **70**, 2051 (1998)
7. Abalos, M., Bayona, J.M., Compañó, R., Granados, M., Leal, C., Prat, M.D.: J. Chromatogr. A **788**, 1 (1997)
8. Serra, H., Nogueira, J.M.F.: J. Chromatogr. A **1094**, 130 (2005)
9. Donard, O.F.X., Randall, L., Rapsomanikis, S., Weber, J.H.: Int. J. Environ. Anal. Chem. **27**, 55 (1986)
10. Craig, P.J., Rapsomanikis, S.: Inorg. Chim. Acta **80**, L19 (1983)
11. Quevauviller, P., Martin, F.M., Belin, C., Donard, O.F.: Appl. Organomet. Chem. **7**, 149 (1993)
12. Song, X., Zapata, A., May, L., Eng, G.: Main Group Chem. **4**, 39 (2005)
13. Eng, G., Desta, D., Biba, E., Song, X., May, L.: Appl. Organomet. Chem. **16**, 67 (2002)
14. Herber, R.H., Stöckler, A., Reichle, W.T.: J. Chem. Phys. **42**, 2447 (1965)
15. Cima, F., Craig, P.J., Harrington, C.: In: Organometallic Compounds in the Environment, p. 101. Wiley, West Sussex, England, (2003)
16. Whalen, D., Lucero, R., May, L., Eng, G.: Appl. Organomet. Chem. **7**, 219 (1993)
17. Williams, M.T.: A history erosion in the Anacostia Drainage Basin. PhD dissertation, The Catholic University of America (1942)

Hyperfine Interact (2006) 170:127–130
DOI 10.1007/s10751-006-9466-3

^{57}Fe Mössbauer spectroscopic study of $(25-x)$MnO–xZnO–15Fe_2O_3–60B_2O_3 glasses

Roshan Lal · S. P. Taneja · C. Wynter

Published online: 9 January 2007

Abstract The ^{57}Fe Mössbauer technique has been used to investigate the effect of zinc oxide substitution in $(25-x)$MnO–xZnO–15Fe_2O_3–60B_2O_3 glass system (x=0, 5, 10, 15 and 20 mol% of ZnO). Mössbauer absorption spectra for all the samples recorded at room temperature suggest the existence of the two paramagnetic quadrupole doublets. The observed variations in hyperfine parameters have been explained on the basis of cations distribution and exchange interaction at the lattice sites and it is concluded that B–B interaction increases while the metal–metal interaction decreases due to replacement of manganese oxide by zinc oxide. These results suggest that the present glass system exhibits a paramagnetic behaviour that changes towards the weak paramagnetic when manganese oxide was replaced with zinc oxide.

Key words Mössbauer technique · zinc oxide substitution · absorption spectra

1 Introduction

^{57}Fe Mössbauer technique is an excellent tool to study the magnetic behaviour in magnetic materials e.g. ferrites, alloys and glasses. Generally glasses are disordered materials having non-periodic arrangement of atoms. Several metal oxides like ZnO, MnO, PbO, CuO, etc. are mixed with B_2O_3 to form glasses.

R. Lal (✉)
Department of Physics, Kurukshetra University, Kurukshetra 136 119, India
e-mail: roshandhiman_kuk@yahoo.co.in

S. P. Taneja
Department of Physics, Maharshi Dayanand University, Rohtak 124 001, India
e-mail: dr_taneja_sp@hotmail.com

C. Wynter
Department of Chemistry, Nassu Community College, Garden City, NY, USA
e-mail: wynterc@ncc.edu

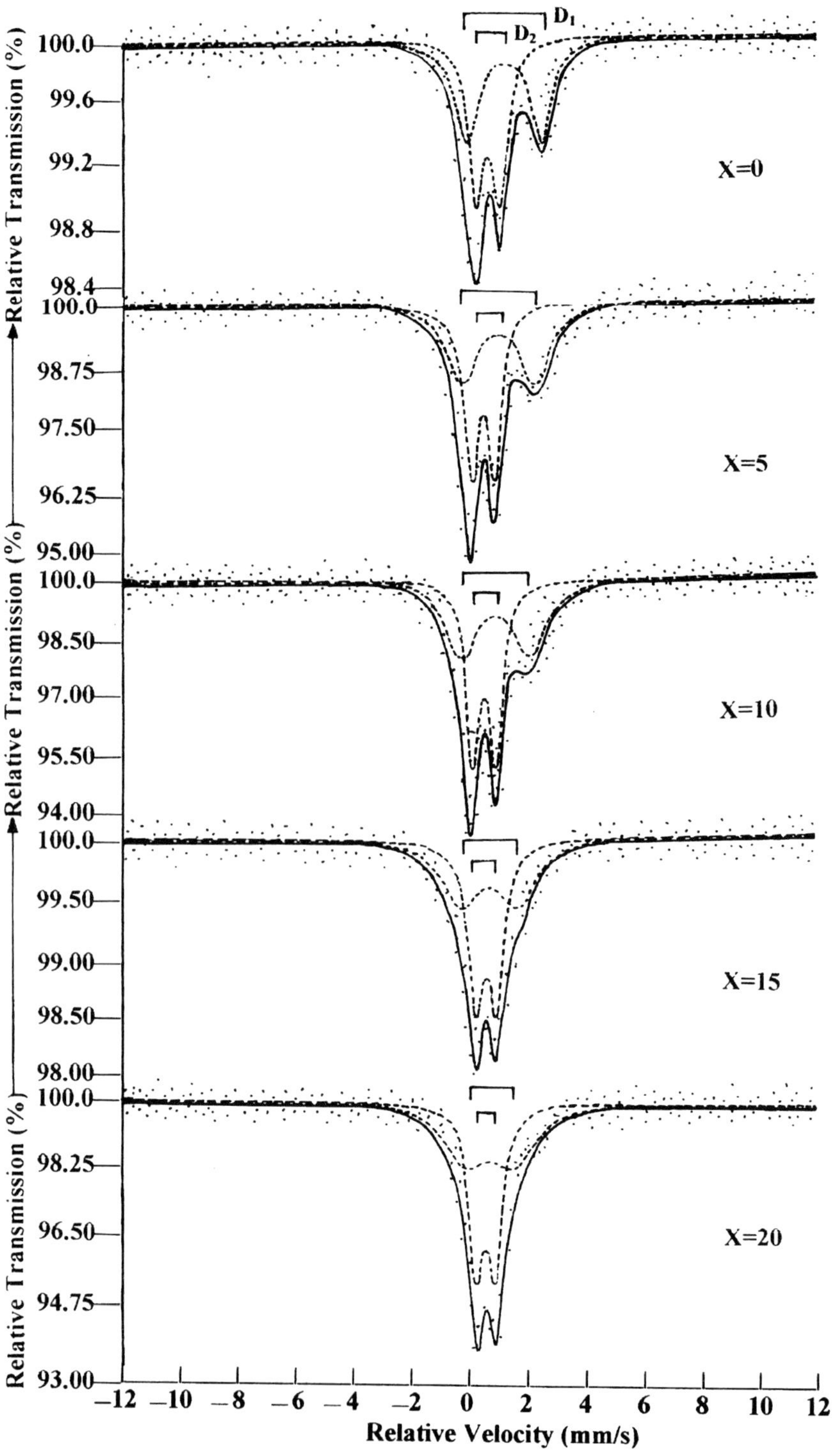

Fig. 1 Mössbauer absorption spectra for X=0, 5, 10, 15 and 20 recorded at room temperature

Table 1 Hyperfine interaction parameters computed from Mössbauer spectra recorded at room temperature,'δ' is given relative to α-Fe

Composition X (mol%)	Site	Isomer shift ($\delta\pm0.01$) mm/s	Quadrupole splitting ($\Delta\pm0.01$) mm/s	Line width ($\Gamma\pm0.01$) mm/s	χ^2/degree of freedom
X=0	D_1	0.87	4.82	0.93	0.39
	D_2	0.37	1.53	0.63	
X=5	D_1	0.78	4.62	1.19	0.32
	D_2	0.23	1.48	0.55	
X=10	D_1	0.58	4.44	1.28	0.33
	D_2	0.23	1.48	0.52	
X=15	D_1	0.33	3.65	1.47	0.43
	D_2	0.25	1.32	0.59	
X=20	D_1	0.38	3.06	1.67	0.31
	D_2	0.29	1.25	0.57	

D_1 and D_2 corresponds to doublet one and two

Recently the study of zinc oxide in the form of zinc ferrite thin film [1, 2] as well as magnetic properties of zinc substituted manganese ($Mn_{1-X}Zn_XFe_{2-X}O_4$) ferrites system [3] received considerable attention due to their structural characteristics and its uses in different commercial applications [4].

When manganese ferrite diluted with zinc then it shows a ferromagnetic to a week paramagnetic behaviour. So we plan to prepare such system in the form of borate glasses in order to study its magnetic behavior.

2 Experimental

Zinc substituted Manganese doped iron borate glasses were prepared from regent grade powders of ZnO, $MnCO_3$, Fe_2O_3 and H_3BO_3 by solid-state reaction technique. The general formula was $(25-x)MnO–xZnO–15Fe_2O_3–60B_2O_3$, here x=0, 5, 10, 15 and 20 in mol% of zinc oxide. The X-ray diffraction patterns shows that all the samples are completely amorphous in nature. Mössbauer absorption spectra of all the samples recorded at room temperature and were fitted to two paramagnetic quadrupole doublets as shown in Fig. 1.

3 Results and discussion

From the experimental observations such as IR and Mössbauer study, it has been observed [5, 6] that no tetrahedral formation occur which suggest that Mn^{2+}(MnO) probably changes to Mn^{3+} at higher temperature.

The variation in observed hyperfine parameters has been explained on the basis of cation distribution and exchange interactions between iron ions and zinc ions at the octahedral (B) and tetrahedral (A) sites. It has been reported that Mn^{3+} has strong preference to occupying only the octahedral B-site, however ZnO (Zn^{2+} ion) has strong chemical preferentially occupy A-site.

When zinc oxide is introduced at the cost of Manganese oxide, there is consistent decrease in hyperfine parameters listed in Table 1. This is due to the fact that some of the zinc oxide is lost (during firing process as zinc oxide is more volatile having lowest values of melting point and vapour pressure than those for other elements) while the remaining zinc ion migrate iron ions from A to B site in view of the site preferences for different ions as mentioned above. This increases the iron concentration at B site. However as zinc oxide concentration increases, the iron ions left at A site being small in number, the A–B interaction experienced by B-site iron ions decreases. Also, the increased number of iron ions at the B site increases the B–B interaction; resulting in spin canting consequently, the hyperfine parameters decrease. Moreover, the possible interactions at the B-site are as follows : (1) Fe^{3+}–Fe^{3+}, (2) Fe^{3+}–Mn^{3+}, (3) Mn^{3+}–Mn^{3+}. Out of these Fe^{+3}–Fe^{3+} and Fe^{3+}–Mn^{3+} interactions are very weak while Mn^{3+}–Mn^{3+} interaction is strong antiferromagnetic. As the manganese oxide concentration is being replaced by zinc oxide concentration due to which metal–metal interaction at the lattice site B decreases which causes the depletion in the hyperfine parameters.

Present study shows that when the manganese doped iron borate glasses are diluted with ZnO, show a paramagnetic to a week paramagnetic behavior. However when this system was formed in the form of ferrite [3], showed a ferromagnetic to a weak paramagnetic behavior when diluted with zinc oxide.

On the basis of these results it is concluded that magnetic behaviour of the system depends upon the method of preparation as well as on the sintering temperature.

4 Conclusion

It is concluded that the present glass system shows a paramagnetic behaviour that decreases to a week paramagnetic behaviour when manganese oxide was replaced by zinc oxide.

References

1. Tanaka, K., Nakaashima, S., Fujita K., Hirao, K.: J Phys Condens Matter **15**, 469 (2003)
2. Roy, S., Basu, S.: Bull Mater Sci **25**(6), 513 (2002)
3. Rath, C., Anand, S., Das, R.P., Sahu, K.K., Kulkarni, S.D., Date. S.K., Misra, S.C..: J Appl Phys **91**(4), 2211 (2002)
4. Wang, Z.L.: J Appl Phys Condens Matter **16**(25), 829 (2004)
5. Lal, R., Sharma, N.D., Sharma, H.K., Chandra, K.: Indian J Pure Appl Phys **42**, 25 (2004)
6. Lal, R., Sharma, N.D., Chandra, K.: Indian J Pure Appl Phys **42**, 498 (2004)

Hyperfine Interact (2006) 170:131–143
DOI 10.1007/s10751-006-9463-6

Studying surfaces and thin films using Mössbauer spectroscopy

Laura K. Perry · D. H. Ryan · R. Gagnon

Published online: 5 December 2006

Abstract Using a series of bi-layer samples, we show how Conversion Electron Mössbauer Spectroscopy (CEMS) and X-ray Backscatter Mössbauer Spectroscopy (XBS) can be done with the same experimental set up. The penetration depths of the K and L conversion electrons are measured as 51(6) and 330(240) nm, respectively, with relative contributions of 88(9) and 12(9)%. The penetration depth of the Fe-K_α X-ray signal is determined to be 3.6(2) μm. As a demonstration we show data on surface damage effects in electropolished TRIP steels, and by comparing CEMS and XBS Mössbauer patterns we estimate the thickness of a damaged layer (created by sanding) to be 550(50) nm.

Key words Mössbauer spectroscopy · CEMS · XBS · Thin films · K conversion electrons · L conversion electrons · X-ray backscattering

1 Introduction

Conventional transmission Mössbauer spectroscopy is used to study bulk sample properties: magnetic, chemical and structural. Resonance is detected by the loss of transmitted intensity, giving an absorption spectrum. Each resonant absorption event creates a nucleus in an excited state, and this nucleus must return to the ground state by re-emitting radiation (γ-ray, X-ray, conversion electron and auger electron). These backscattered radiations are used in conversion electron Mössbauer spectroscopy (CEMS) and X-ray backscattering (XBS).

CEMS and XBS are backscattering methods in which the detected signal comes from the incident face of the sample. A backscattering approach is attractive if

L. K. Perry (✉) · D. H. Ryan · R. Gagnon
Center for the Physics of Materials, Department of Physics,
McGill University, Rutherford Building, 3600 University Street,
Montréal, Québec, Canada H3A 2T8
e-mail: perryl@physics.mcgill.ca

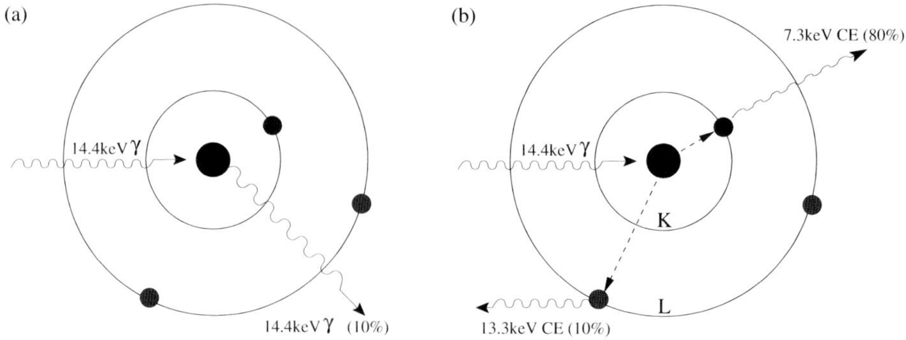

Figure 1 Internal processing of incident γ-radiation in ^{57}Fe. The nucleus may **a** re-emit a 14.4 keV γ-ray (10%) or **b** transfer the energy to shell electrons through virtual photons, emitting K-level (7.3 keV at 80%) and L-level (13.3 keV at 10%) conversion electrons [4].

the sample is either too thick (where the transmission signal would be completely absorbed), too thin (where absorption is negligible), or where the surface is the focus of the investigation. The penetration depths of conversion electrons and X-rays are much lower than γ-rays, and thus the backscattered signal must come from events close to the surface in order to be detected. This makes both CEMS and XBS surface-biased techniques, yielding Mössbauer spectra with all of the conventional magnetic, chemical and structural information, but derived from nuclei close to the sample surface. Because conversion electrons are less penetrating than X-rays, CEMS is more surface-biased than XBS and is the conventional method chosen to study surfaces and thin films. X-rays can escape from deeper within the sample and thus give more information on bulk sample properties.

Combining CEMS and XBS data on the same sample allows surface (~120 nm) and deeper (~8,100 nm) contributions to be distinguished. In this study, we examine bi-layered samples consisting of a thin iron overlayer (ranging between 17(1) nm and 6.1 μm) on a 304 stainless steel substrate, through CEMS and XBS. The difference between the Mössbauer patterns of non-magnetic stainless steel and magnetic iron motivates this choice of sample, because it allows the CEMS and XBS signals arising from each material to be distinguished. As an example of application, we show data on surface damage effects in electropolished TRIP steel, an alloy with magnetic and non-magnetic phases and thus whose Mössbauer pattern resembles that of the bi-layered samples. Through CEMS and XBS, the surface and bulk of TRIP steel can be studied, and since the surface is very sensitive to damage, comparing the two methods can provide insight into optimal surface preparation techniques. The penetration depths of conversion electrons and X-rays in iron are used to determine the thickness of a damaged layer created by sanding a TRIP steel surface.

2 Conversion electron Mössbauer spectroscopy

When incoming γ-radiation excites ^{57}Fe to its first excited state at 14.4 keV, the nucleus can shed the extra energy through several means. One of these processes, shown in Figure 1a, is the re-emission of a 14.4 keV γ-ray (10% probable). This

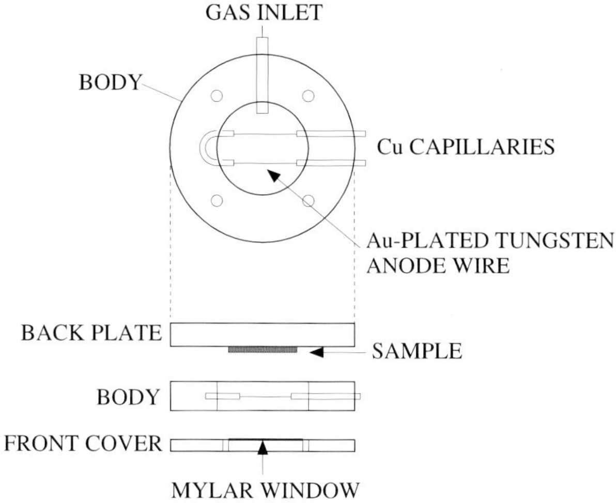

Figure 2 Sectional views of the CEMS detector. The *top diagram* illustrates a section through the anode plane, and the *bottom* shows a side view of the separate components.

backscattered γ-radiation is used in selective excitation double Mössbauer spectroscopy (SEDM) [1, 2], and can also be used as another means of backscattering Mössbauer spectroscopy (instead of CEMS and XBS) [3]. The more probable process is shown in Figure 1b, in which the nucleus transfers the energy to a 1*s* or *K* shell electron through a virtual photon, resulting in the emission of a 7.3 keV conversion electron (80% probable), or to a 2*s* or *L* shell electron, emitting a 13.3 keV conversion electron (10% probable). The energy of the outgoing radiation is given by $E_{CE} = E_\gamma - E_B$, where E_B is the binding energy of that shell. All probabilities are from [4].

Conversion electrons are detected through Conversion Electron Mössbauer Spectroscopy (CEMS [5]). A Mössbauer spectrum arises because the backscattered radiation is new radiation emitted at resonance. As the 7.3 keV electrons penetrate through very little matter, they are only detected when they are emitted near the surface of a sample. CEMS is most often chosen because of this bias to the sample surface, and is commonly used on thin films and treated surfaces [6, 7], as well as in corrosion studies. Since CEMS is a backscattering method, it is also a possible approach when a sample is too thick to be evaluated through transmission Mössbauer spectroscopy, where the beam is entirely absorbed in the thick material.

2.1 CEMS detector design

Since conversion electrons have shallow penetration depths (90% of the *K*-conversion electron signal comes from the first ~120 nm of the sample), there can be no window between the sample and detector, as any window material would stop the electrons before they could enter the detector. Most events occur too deep to be detected, so the count rate is low. In addition, those events occurring near the surface often are not detected because the electron emission is in the wrong direction. Detector efficiency is therefore key in the design of the CEMS detector.

The CEMS detector shown in Figure 2 is a continuous-flow gas-filled proportional counter, similar in design to the detector built by Liao et al. [8]. Alternate CEMS detector designs are given by [5, 9]. The detector has three main parts: a front

Figure 3 Conversion electron Mössbauer spectra of Fe/SS for a range of Fe overlayer thicknesses of 17(1) to 368(1) nm. At 368(1) nm, the magnetic peaks of Fe dominate the spectrum.

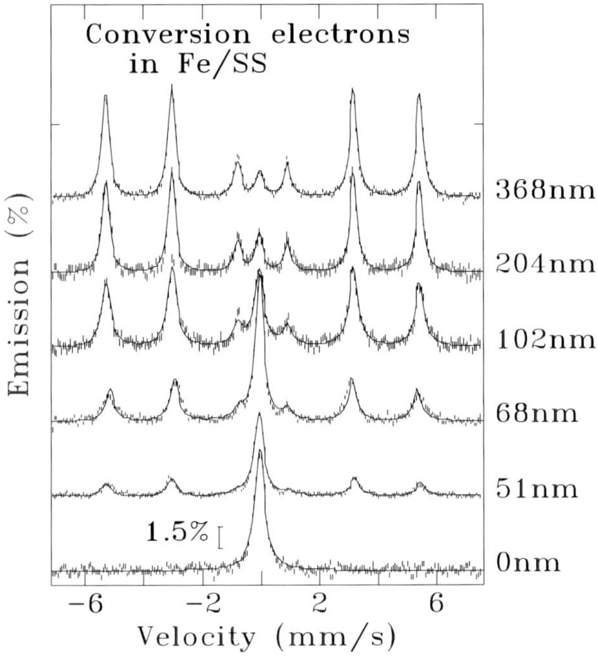

window (0.7 mm thick Plexiglass), a body and a backplate, and all of the inner surfaces are coated with aluminum and connected to ground. The ^{57}Co Mössbauer source emits not only 14.4 keV γ-rays, but also 6.4 keV Fe-K_α X-rays, which produce a background signal of photoelectrons. Photoelectrons are also produced when non-resonant 14.4 keV γ-rays and higher energy photons scatter from non-resonant materials [8]. The detector body is thus made of Plexiglass, because low Z materials help to reduce photoelectron production. The body carries the anode wire, the high voltage connection, and the fill gas line. The 25 μm anode wire is made of tungsten, for strength under tension, and plated with gold to avoid corrosion and contamination. A voltage of ~1000 V is applied to the anode. Lower voltages produce lower gain, while higher voltages cause breakdown. The sample is mounted on the backplate, with the sample surface directly exposed to the gas flow [5]. The fill gas used in the proportional counter is 4%CH_4/He. Helium is chosen as the primary gas because it is transparent to most radiation except conversion electrons. For instance, at atmospheric pressure, a 3.5 mm thick He gas layer has efficiencies of less than 0.02% for 6.4 keV X-rays and less than 0.002% for 14.4 keV γ-rays [10]. The fill gas is pre-mixed with CH_4, a quench gas that is present to absorb de-excitation photons emitted when collisions between electrons and neutral gas atoms excite the gas atoms. It thus stabilizes the proportional region of the detector. The flow rate was approximately 30 ml/min.

2.2 Conversion electron ranges in Fe

Mössbauer spectra were collected using a 3.7 GBq Rh ^{57}Co source mounted on a constant-acceleration spectrometer, calibrated using α-Fe foil at room temperature.

Figure 4 Natural logarithm of the stainless steel signal vs. Fe overlayer thickness for conversion electrons. The K-conversion electrons are attenuated by shallower overlayer thicknesses than L-conversion electron. This introduces a break in slope where the L-conversion electron signal starts to dominate over that of the K-conversion electrons.

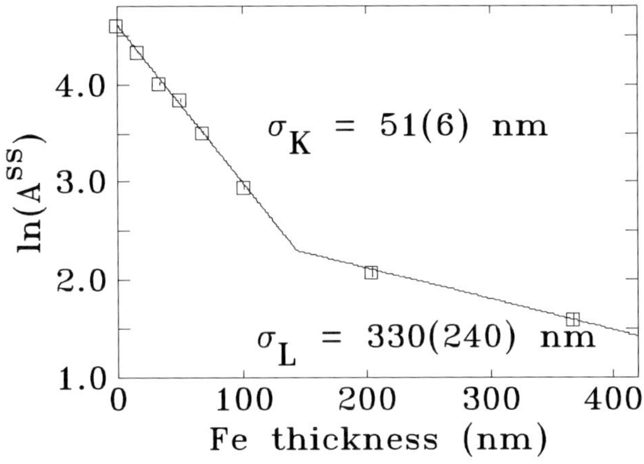

Table I Binding energies, conversion electron energies, penetration depths and emission probabilities of K- and L-conversion electrons in Fe/SS (from CEMS)

Shell	E_B (keV)	E_{CE} (keV)	σ (nm)	% Probability
K	7.1	7.3	51(6)	88(9)
L	0.7	13.3	330(240)	12(9)

Typical count rates were 300 counts per second. The iron films were deposited at room temperature by DC-magnetron sputtering onto non-magnetic 304 stainless steel substrates. We used an argon pressure of 1 x 10^{-2} torr and a power of 50 W, leading to a deposition rate of around 5 nm/min. The overlayer thicknesses were estimated using this deposition rate, then determined by X-ray reflectivity. The iron film thicknesses varied from 17(1) to 368(1) nm.

Figure 3 shows the conversion electron Mössbauer spectra of Fe/SS for overlayer thicknesses ranging between 17(1) and 368(1) nm. The central stainless steel peak rapidly decreases in intensity as the overlayer thickness increases. This is due to a shift in balance between the substrate and overlayer signals, where at thick overlayers, the iron signal dominates. The stainless steel conversion electron signal decreases as $A^{ss} = A_o^{ss} e^{-d/\sigma}$, where A_o^{ss} is the initial stainless steel area for zero overlayer thickness, A^{ss} the measured stainless steel area, d the overlayer thickness, and σ the penetration depth of conversion electrons in iron. Figure 4 shows the natural log of A^{ss} as a function of overlayer thickness, and the inverse of the slope gives the penetration depth. The rapidly attenuated part is due to the K-conversion electrons (7.3 keV), which have a lower penetration depth than L-conversion electrons (13.3 keV) but a higher emission probability. The slowly attenuated part, due to the L-conversion electrons, dominates at thicknesses greater than 200 nm because the K-conversion electrons are greatly attenuated by the thicker overlayers. The measured penetration depths are 51(6) nm for K-conversion electrons and 330(240) nm for L-conversion electrons (Table I).

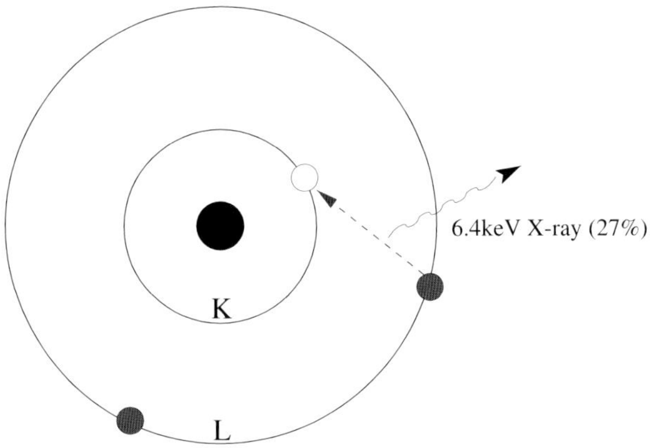

Figure 5 An L shell electron falls to fill the hole on the K shell introduced by the K-conversion electron emission. This creates a 6.4 keV X-ray (27%), or a 5.6 keV Auger electron (63%) [4].

3 X-ray backscattering

When the nucleus is excited by incident 14.4 keV γ-radiation, the energy can be transferred to a K shell electron by means of a virtual photon, emitting a K-conversion electron (Section 2). The void created by the loss of the K shell electron is filled by an L shell electron, and the decay from higher to lower shell causes the emission of an X-ray with an energy corresponding to the difference in binding energy of the electronic shells (Figure 5). In the case of ^{57}Fe, the difference in binding energies of the K and L shells is 6.4 keV, which is the Fe-K_α that is used in X-ray backscattering (XBS).

The main difference between CEMS and XBS is that X-rays are far more penetrating than conversion electrons, so that XBS is not as much a surface-sensitive technique. Due to the greater range of X-rays, the space between sample and detector is not as much a concern as in the case of CEMS. This allows for more liberty in sample location, because although the sample can be mounted in the detector chamber, it can also be located outside the detector chamber without losing the X-ray signal. Overall, XBS yields all the same information about a sample as would CEMS, but with lesser surface bias (90% of the XBS signal comes from the first ~8,100 nm of the sample, instead of ~120 nm for conversion electrons).

3.1 XBS detector design

The initial design of an Fe-K_α X-ray detector used a forward-scattering geometry which allowed the scattered X-rays to be detected in a solid angle of up to 2π [11]. The first appearance of an XBS detector was given by Swanson et al. [5] 3 years later in 1970. The XBS detector was then later designed to also count both X-rays and conversion electrons simultaneously [12], and eventually transmission γ-rays are also included [10, 13]. The detector used for XBS in this study is the same continuous-flow gas-filled proportional counter as that described in Section 2.1, save for a few details. The biggest concern are the Fe-K_α X-rays incident from the source. The source emits even more X-rays than γ-rays (Figure 6a), and being that the detector is now sensitive to X-rays, a proper filter must be chosen. A Plexiglass filter 3–4 mm thick nearly eliminates the incident X-ray intensity while only slightly reducing the

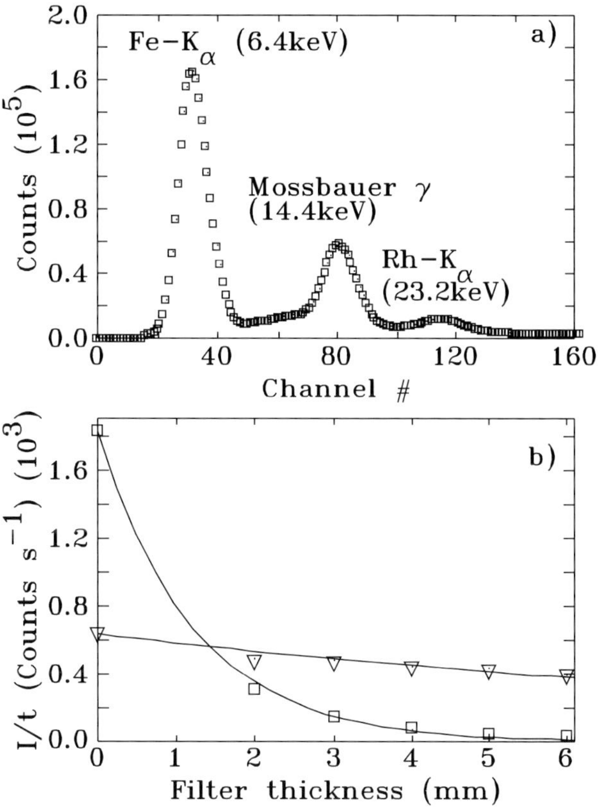

Figure 6 **a** Pulse height pattern of radiations emitted from the ^{57}Co source. There are 6.4 keV Fe-K_α X-rays, 14.4 keV γ-rays, and 23.2 keV Rh-K_α X-rays. **b** X-ray and γ-ray intensity (divided by time) for varying Plexiglass filter thicknesses.

γ-ray intensity (Figure 6b). The fill gas used is 10%CH_4/Ar. The absorption depth of 6.4 keV X-rays in Ar is on the order of 3 cm, and that of 14.4 keV γ-rays is 25 cm [14]. The primary component of the fill gas is therefore chosen to be Ar because it has a higher X-ray stopping power than He, while being transparent to the 14.4 keV γ-rays. The flow rate was approximately 30 ml/min.

A side view of the CEMS/XBS detector is shown in part (a) of Figure 7, with the sample mounted inside the detector chamber. Since Ar has a higher stopping power than He, both X-rays *and* conversion electrons are detected in the detector chamber. The detector energy window set by the signal channel analyzer (SCA) is narrowed around the 6.4 keV X-ray. The *K*-conversion electrons (7.3 keV) emitted by the sample are similar in energy to X-rays, and they too are detected in this energy range and contribute approximately 20% to the overall signal. In part (b) of Figure 7, the detector's backplate was replaced by an aluminized Mylar window. The sample was mounted behind the Mylar window, outside of the detector chamber. This eliminated the conversion electron signal that appeared in the former setup.

3.2 X-ray range in iron

In addition to the sputtered samples described in Section 2.2, iron foil was used as an added overlayer. The iron foil was made by Goodfellow and is 99.85% pure, with

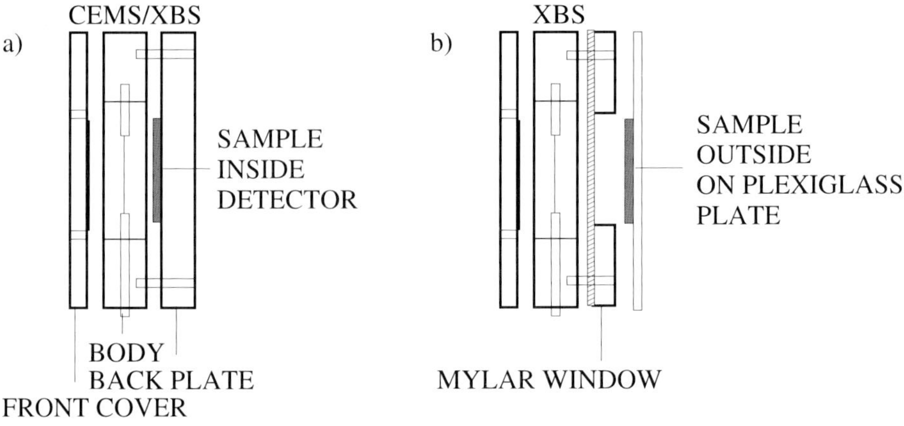

Figure 7 **a** Side view of the CEMS/XBS detector, with the sample mounted inside the detector chamber. **b** Side view of the XBS detector with the sample mounted outside to eliminate contaminant conversion electrons.

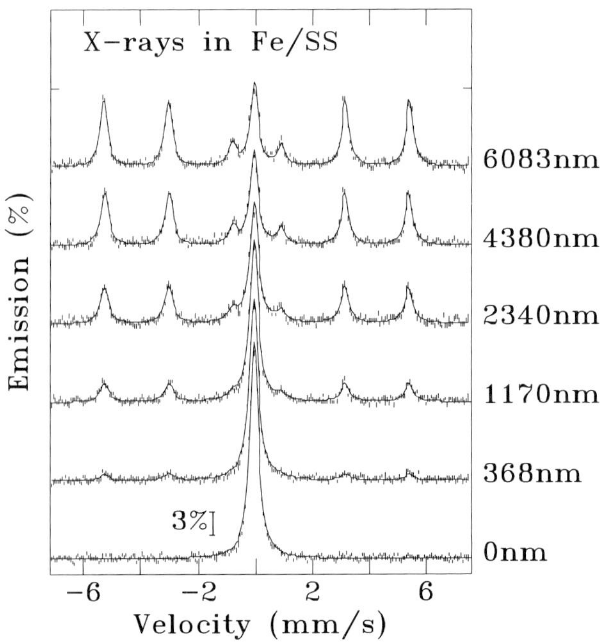

Figure 8 X-ray backscattering Mössbauer spectra of Fe/SS for a range of Fe overlayer thicknesses of 368(1) to 6083 nm. To reduce the stainless steel spectral area, the overlayer thickness must be larger than that required to attenuate the less penetrating conversion electrons.

dimensions 50 × 50 × 0.001 mm. The overlayer thicknesses varied from 17(1) nm to 6.1 μm. Thicker overlayers are needed due to the greater range of Fe-K_α X-rays. Typical count rates were 100 counts per second.

Figure 8 shows XBS Mössbauer spectra, with Fe overlayer thicknesses ranging between 368(1) nm and 6.1 μm. On comparison with Figure 3, we immediately see that a greater overlayer thickness is required to attenuate the XBS signal from stainless steel. Even at 368 nm, the magnetic Fe peaks are only just beginning to

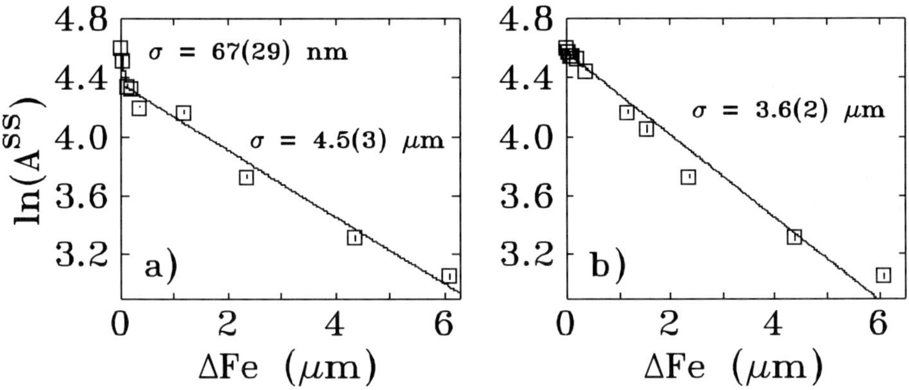

Figure 9 Natural logarithm of the percent stainless steel signal vs. Fe overlayer thickness (ΔFe) for X-rays, with overlayer thicknesses ranging between 17(1) nm and 6.1 μm. **a** The sample is mounted inside the detector chamber, and the signal receives contributions from 6.4 keV X-rays (76(3)%) as well as surface conversion electrons (24(3)%). **b** The sample is mounted outside of the detector chamber, and the aluminized Mylar window between the sample and detector eliminates the conversion electron signal.

Table II Penetration depths and relative signal contributions of K-conversion electrons and X-rays in Fe/SS, as obtained through XBS and with the sample mounted inside and outside of the detector chamber. The % column shows the relative distribution of the different signals

RADIATION	Inside σ	%	Outside σ
K-level conversion e^-	67(29) nm	24(3)	-
X-ray	4.5(3) μm	76(3)	3.6(2) μm

appear. This illustrates the greater penetrating power of X-rays as compared to conversion electrons.

Figure 9 shows the natural log of the stainless steel signal area (A^{SS}) for overlayer thicknesses ranging between 17(1) nm and 6.1 μm. In part (a) of Figure 9, the sample was mounted inside the detector (Figure 7a), and a break in slope appears at ~90 nm. The penetration depth obtained from the most rapidly attenuated contribution is 67(29) nm, which is within error of the K-conversion electron range of 51(6) nm determined in Section 2.2. For part (b) of Figure 9, the sample was mounted outside the detector, and an aluminized Mylar window stops the K-conversion electrons from entering the detector (Figure 7b) so that only the X-rays contribute. The penetration depth of X-rays in iron was determined to be 3.6(2) μm. Table II summarizes the results for XBS.

4 Application to TRIP steel

When an austenitic steel is plastically deformed at room temperature, some of the austenitic phase (solid solution of carbon in FCC or γ-Fe) in the most severely strained portions of the steel will transform to martensite (BCC form of iron in which some carbon is dissolved). TRIP is an acronym for the **TR**ansformation **I**nduced

Plasticity of a metal alloy, and the strain-induced decomposition of the austenitic phase is an important contributor to the observed properties of the alloy [15]. The term *retained austenite* refers to the stabilization of residual austenite by the carbon that is partitioned from bainitic ferrite [15]. A typical phase composition consists of 10% retained austenite (face-centered cubic), 20% bainitic ferrite (body-centered cubic) and balance allotriomorphic ferrite [16]. In short, TRIP steel consists of small austenitic inclusions in a ferrite matrix.

The distinguishing characteristic of the FCC austenitic phases is that they are non-magnetic. The Mössbauer pattern of austenite is thus a single peak, whereas that from the ferritic phases consists of six peaks, typical of magnetic iron-bearing alloys. This closely resembles the bi-layered samples examined by CEMS and XBS in Sections 2 and 3, where the Mössbauer patterns consisted of a single peak for stainless steel and six peaks for magnetic iron.

A common approach to determining the amount of retained austenite in a TRIP steel is Cu-K_α X-ray diffraction (XRD). When a sample is evaluated through XRD, even at normal incidence the X-ray signal from a depth d in the sample must make two passes through that thickness prior to detection (one upon incidence and another upon escape), so that the overall travel path length is $2d$. In contrast, the XBS X-ray originates from within the sample itself, and makes only a single pass through d before escaping. Furthermore, the 6.4 keV Fe-K_α and the 8.05 keV Cu-K_α X-rays lie on either side of the iron K edge at 7.11 keV. Considering the total attenuation of the Cu-K_α X-ray (297.80 cm^2/g compared to 70.34 cm^2/g for Fe-K_α X-rays) and the amount of material through which it must pass for detection, Cu-K_α X-rays are least eight times more attenuated by iron than Fe-K_α X-rays. Consequently, the majority of Cu-K_α XRD signals come from the surface of the sample, and as the TRIP steel surface is highly sensitive to damage, XRD may not be as reliable a technique for determining the bulk retained austenite fraction as Fe-K_α XBS.

TRIP steels are prone to surface damage because the austenitic phases readily transform to martensite under stress. Samples for phase analysis are often cut from larger billets, therefore any method which probes only surface properties, such as Cu-K_α X-ray diffraction and CEMS, will underestimate the retained austenite fraction because the surface phases are destroyed. To a certain extent, the damaged surface can be removed through electropolishing, a chemical etch intended to remove the damaged surface. However, bulk methods such as neutron diffraction and XBS are preferred because they depend less on the efficiency of surface preparation techniques. By comparing CEMS and XBS on the same TRIP steel sample, we explore surface damage effects and demonstrate that even simple mechanical working suffices to destroy the surface austenitic phases. Through XBS and CEMS measurements we obtain an estimate for the thickness of the damaged layer created through sanding.

4.1 Results

The CEMS and XBS Mössbauer spectra of an electropolished TRIP steel (with a ~7% retained austenite content) are shown in Figure 10 (left). They were fitted with three components: two magnetic (ferrite and martensite) and one non-magnetic (austenite). The difference in retained austenite fraction measured by CEMS and XBS (3.4(6) and 6.4(6)%, respectively, Table III) suggests that the damaged surface

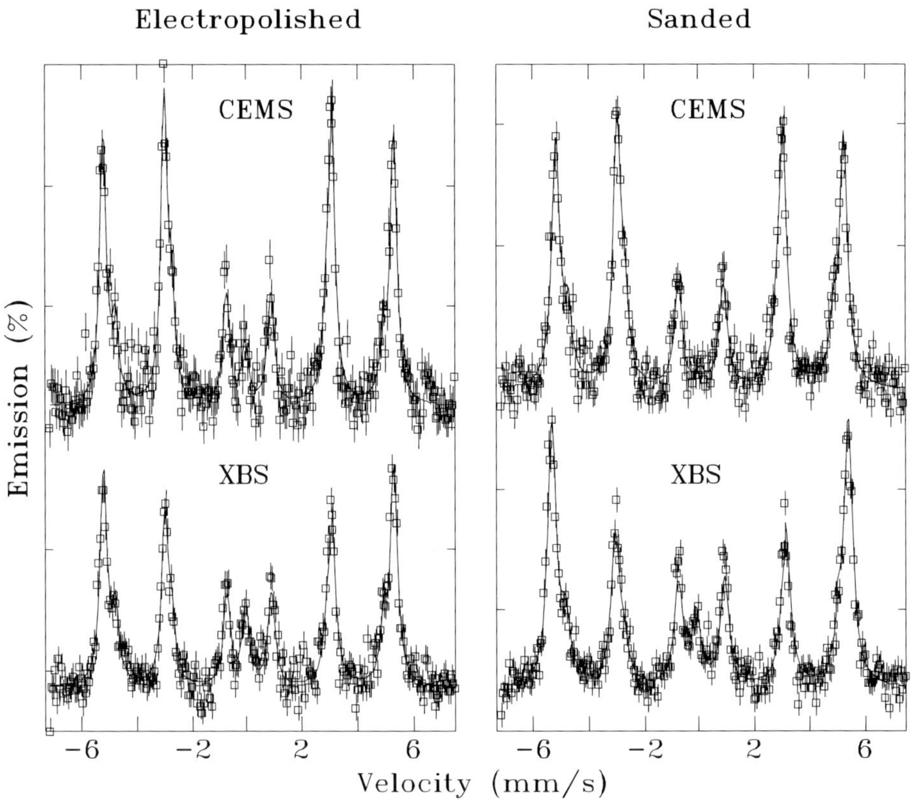

Figure 10 Conversion electron and X-ray backscattering Mössbauer spectra for TRIP steel, with electropolished and sanded surface.

has not been completely removed through electropolishing. Cu-K_α X-ray diffraction would give an even lower estimate of austenite content than CEMS, emphasizing the importance of the efficiency of surface preparation if surface-biased probes are used.

When the electropolished surface is sanded (600 grit sanding paper), a new damaged layer is created. The stress of sanding transforms the surface austenitic phases to martensite. Figure 10 (right) shows the CEMS and XBS Mössbauer spectra of the TRIP steel with the damaged surface. While the central austenitic component is present in the electropolished CEMS and XBS patterns (left), it survives only in the XBS pattern of the sanded TRIP steel (right). The central peak disappears completely in the right CEMS pattern, indicating that the surface is completely destroyed and that the damaged layer is at least 100 nm thick. The XBS patterns show only a 14% loss of austenite signal in going from electropolished to sanded surfaces, reinforcing this choice of method for the measurement of retained austenite fraction.

Table III lists the retained austenite content of the TRIP steel, from CEMS and XBS. In order to estimate the thickness of the damaged layer, we model the sanded TRIP steel as a bi-layered sample, with a damaged austenite-free surface layer overlaying bulk TRIP steel. The retained austenite fraction of the underlayer is estimated from XBS measurements on the electropolished surface. There are

Table III Conversion electrons and X-rays emitted from austenitic phases in TRIP steel, with electropolished and sanded surface. A^{aus} is the measured area of the austenitic phase (identical to A^{ss} from Sections 2.2 and 3.2). The conversion electron austenite signal is completely eliminated by the damaged layer created through sanding, and the X-ray signal is only slightly reduced

RADIATION	Setup	A^{aus} (%)	
		Electropolish	Sanded
X-rays	Inside	6.4(6)	5.1(9)
	Outside	-	5.5(5)
Conversion e$^-$'s	Inside	3.4(6)	0.9(5)

therefore two contributions to the overall signal: one from the bulk (6.4(6)% retained austenite), and another from the damaged surface (0% retained austenite). The overall XBS retained austenite fraction from both contributions is 5.5(5)% (Table III). The measured austenite area A^{aus} is given by $A^{aus} = A_o^{aus} e^{-d/\sigma}$, where A_o^{aus} is the initial austenite area in the absence of the damaged layer. The damaged layer thickness d can be obtained by integrating over the two contributions to A^{aus}:

$$x_o = x_d \int_0^d A_o^{aus} e^{-x/\sigma} + x_b \int_d^\infty A_o^{aus} e^{-x/\sigma} \tag{1}$$

where x_o is the overall retained austenite fraction (5.5(5)%), x_d is zero (the damaged layer contains no austenite), x_b is the retained austenite fraction of the bulk (6.4(6)%), and σ the penetration depth of X-rays in iron (3.6(2) μm). From this we obtain the damaged layer thickness d = 550(50) nm.

5 Conclusions

Through conversion electron Mössbauer spectroscopy and X-ray backscattering, we have determined penetration depths for conversion electrons and X-rays in iron. Both methods are easy and can be performed using the same detector, with only minor changes. For CEMS, the detector can distinguish between conversion electron contributions from different electronic shells, and for XBS, the detector can be tuned to detect only X-ray emissions. CEMS and XBS are both useful in determining the retained austenite content of a TRIP steel, and the comparison between the measured retained austenite fraction obtained by the two methods demonstrates the extreme surface sensitivity of the alloy. Simple mechanical working such as sanding is sufficient to create a damaged layer that is half a micron thick. Electropolishing does remove some of the surface damage, but it is preferable to use bulk methods such as Neutron scattering and XBS if the purpose of the study is to determine the retained austenite fraction.

References

1. Van Lierop, J., Ryan, D.H.: Rev. Sci. Instrum. **72**, 3349 (2001)
2. Van Lierop, J., Ryan, D.H.: Hyperfine Interact. **141/142**, 141 (2002)
3. Keisch, B.: Nucl. Instrum. Methods **104**, 237 (1972)

4. Liljequist, D.: Mössbauer spectroscopy with electrons. Scanning Electron Microsc. **3**, 997–1017 (1983)
5. Swanson, K.R., Spikerman, J.J.: J. App. Phys. **41**, 3155 (1970)
6. Schaaf, P., Bauer, P., Gonser, U.: Z. Metallkunde **80**, 77 (1989)
7. Bibicu, I., Rogalski, M.S., Nicolescu, G.: Phys. Status Solidi, B. **178**, 459 (1993)
8. Liao, L.X., Ryan, D.H., Altounian, Z.: Rev. Sci. Instrum. **64**, 679 (1993)
9. Bonchev, Z.W., Jordanov, A., Minkova, A.: Nucl. Instrum. Methods **70**, 36 (1969)
10. Bibicu, I., Rogalski, M.S., Nicolescu, G.: Meas. Sci. Technol. **7**, 113 (1996)
11. Hershkowitz, N., Walker, J.C.: Nucl. Instrum. Methods **53**, 273 (1967)
12. Isozumi, Y., Lee, D.-I., Kádár, I.: Nucl. Instrum. Methods **120**, 23 (1974)
13. Bibicu, I., Rogalski, M.S., Nicolescu, G.: Nucl. Instrum. Methods, B. **94**, 330 (1994)
14. McMaster, W.H., Del Grande, N.K., Mallett, J.H., Hubbell, J.H.: Compilation of X-ray cross sections, Sec. 2, Rev. 1. U.S. Atomic Energy Commission Report UCRL-50174, pp. 350. University of California, Livermore, USA (1969)
15. Sheriff, M.Y., Garcia Mateo, C., Sourmail, T., Bhadeshia, H.K.D.H.: Mater. Sci. Technol. **20**, 319 (2004)
16. Bhadeshia, H.K.D.H.: ISIJ Int. **42**, 1059 (2002)

Hyperfine Interact (2006) 170:145–158
DOI 10.1007/s10751-006-9464-5

Variations of $BaSnF_4$ fast ion conductor with the method of preparation and temperature

Georges Dénès · Jamil Hantash · Abdualhafeed Muntasar · Philip Oldfield · Alan Bartlett

Published online: 30 January 2007

Abstract Ionic conductors are solids that have a large number of defects and easy pathways that make it possible for ions to move over long distances in an electric field. In order to be mobile an ion must be small and have a low charge. The fluoride ion is the most mobile anion. The highest performance fluoride ion conductors contain divalent tin, and have a highly layered crystal structure related to the CaF_2 fluorite type. $BaSnF_4$ has the α-$PbSnF_4$ structure, which is a $\sqrt{2}/2\times\sqrt{2}/2\times2$ superstructure of the fluorite type, where the tetragonal unit-cell and the value of the *a* and *b* parameters being equal to half the diagonals of the *(a,b)* face of fluorite are due to the loss of the F Bravais lattice, and the Sn Sn Ba Ba order along the *c* parameter is at the origin of the doubling of the *c* parameter. The $BaSnF_4$ material was prepared first by Dénès et al. (*C. R. Acad. Paris C*, 280: 831, 1975), and its superionic properties were characterized by Dénès et al. (*Solid State Ion.*, 13: 213, 1984). It was found to have a conductivity three orders of magnitude higher than that of BaF_2, with an ionic conduction rate $\tau_i>0.99$. No $BaSnF_4$ was obtained by the aqueous medium, when aqueous solutions of SnF_2 and $Ba(NO_3)_2$ are mixed together; $BaSn_2F_6$ was obtained instead. In a new development of this work, $BaSnF_4$ has been obtained by the wet method for the first time. X-ray powder diffraction showed that the $BaSnF_4$ phase obtained by the wet method varies substantially from one sample to another: (a) signification variations of the *c* parameter of the tetragonal unit-cell reveals that the interlayer distance is sensitive to the leaching conditions, possibly because some of the leached ions remain in the interlayer spacing; (b) large variations of the crystallite dimensions and, as a result of the two-dimensionality of the structure, a strong crystallite dimension anisotropy are observed, with $d\| < d\perp$, where $d\|$ and $d\perp$ are the crystallite dimensions parallel to the *four-fold* main axis, and perpendicular to it, respectively, showing that the layers are very thin and the interlayer interactions are very weak. Variable temperature Mössbauer spectroscopy showed an

G. Dénès (✉) · J. Hantash · A. Muntasar
Laboratory of Solid State Chemistry and Mössbauer Spectroscopy,
Department of Chemistry and Biochemistry, Concordia University, Montréal, QC, Canada
e-mail: gdenes@alcor.concordia.ca

J. Hantash · P. Oldfield · A. Bartlett
Charles River Laboratories Montreal Inc., Senneville, QC H9X 3R3, Canada

unusual large variation of the quadrupole splitting with temperature. A tentative explanation based on unusually large bond angles has been proposed.

Key words ionic conductors · fluorides · tin · Mössbauer spectroscopy · synthetic methods · X-ray diffraction

1 Introduction

Ionic conductivity in solids is due to ionic motion at temperatures well below their melting point, coupled with unusually low values of activation energy. In recent years there has been a great deal of experimental and theoretical work performed on the ionic mobility in materials which crystallize in the fluorite type structure, or closely related to it. Most of the tin (II) containing fluoride ion conductors cannot be obtained in the form of useable crystals because the crystallite shape is usually highly anisotropic (sheet or needle-shape crystals) due to the stereoactivity of tin (II) lone pair, which creates very effective cleavage planes. Such materials have potentials for use for thermodynamic measurement devices, specific electrodes, gas sensors, and solid state batteries. $PbSnF_4$, the highest performance fluoride ion conductor, has been used in the fabrication of an oxygen sensor.

Donaldson and Senior in 1967 reported the immediate precipitation of crystalline metal (II) tin(II) fluoride, $SrSn_2F_6$, $BaSn_2F_6$ and $PbSnF_4$ [1]. For the calcium nitrate/tin (II) fluoride system, the same authors found that the precipitate has a variable composition and is semi-amorphous, however, in a subsequent study, Dénès et al. [2] found that, depending on the conditions of preparation, crystalline $CaSn_2F_6$ or a nanocrocrystalline $Ca_{1-x}Sn_xF_2$ fluorite-type solid solution precipitates, or often, a mixture of both. Two new $MSnF_4$ compounds [M=Sr, and Ba] were prepared by Dénès et al. [3] by direct reaction, at 550°C, and furthermore, $PbSnF_4$ was prepared by direct reaction at 250°C. The three $MSnF_4$ compounds were shown to be isotopic and to be a tetragonal distortion of the fluorite-type CaF_2, with M/Sn order along $\vec{c}$, which also makes their crystal structure related to the PbClF type. Among the MF_2 fluorides, those crystallizing in the fluorite type-structure have the highest fluoride ion conductivity. It is presumably the availability of a large number of potential interstitial sites in the middle of the empty F_8 cubes, i.e. those that do not contain a metal ion, which makes easy the formation of Frenkel defects. However, in a recent study, Dénès has shown that this theory does not explain why the conductivity of β-PbF_2 is much higher than that of BaF_2, since it should be easier to move fluoride ions within the BaF_2 lattice, owing to the large size of Ba^{+2} compared to Pb^{+2}, which results in larger empty F_8 cubes. Dénès [4] hypothesized that the presence of the soft ion Pb^{+2} in the typically ionic fluorite structure seems to favor fluoride ion disorder.

The crystal structure of $BaSnF_4$ was solved by using a combination of X-ray and neutron diffractions, EXAFS and ^{119}Sn Mössbauer spectroscopy [5, 6]. It is closely related to the fluorite-type from with some differences as seen in Figs. 1 and 2. The disappearance of the fluoride ion layer that should be between the adjacent tin layers, and its replacement by a sheet of stereoactive lone pairs, create a very efficient cleavage plane which make $BaSnF_4$ a two-dimensional structure that has very strong anisotropic behavior, in contrast with the three-dimensional structure of BaF_2, the latter having isotropic properties. With the fluorine axially bonding tin occupying the F_8 cubes that are empty in BaF_2, the $BaSnF_4$ structure does not offer the same potential interstitial sites for the formation of Frenkel defects as BaF_2 does, and therefore, its ionic conductivity would be expected to be much lower. To the contrary, its fluoride ion mobility is about three orders of magnitude as high as that of BaF_2

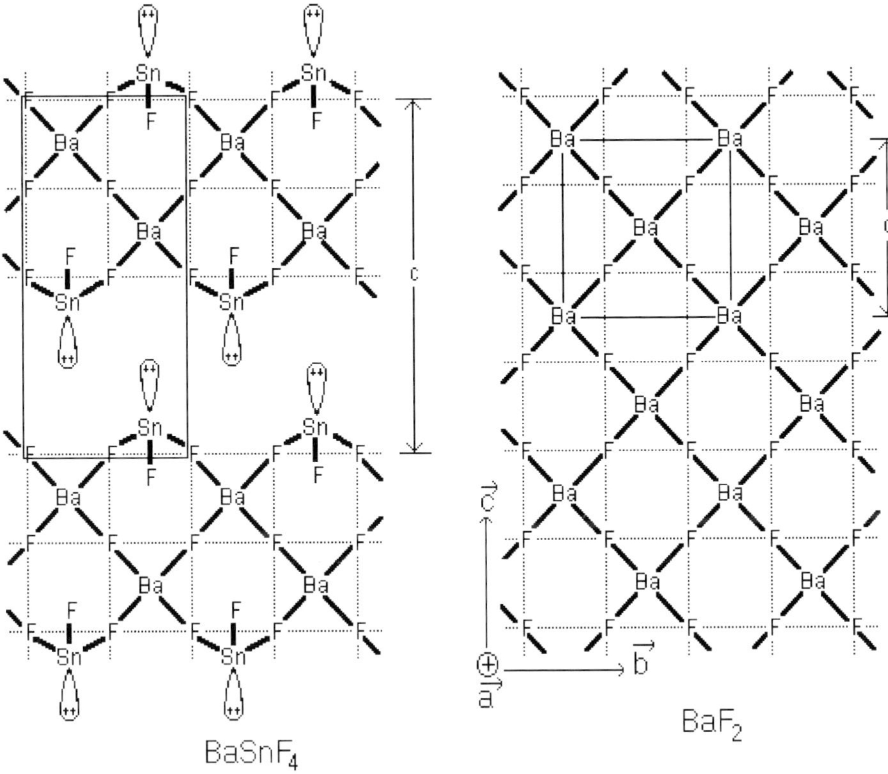

Fig. 1 Projection of a slice of the structure of BaF_2 and $BaSnF_4$ on the (b, c) plane in the BaF_2 axes

[7]. This has been attributed to the fluoride disorder created by the Sn (II) soft atom which is covalently bonded in an ionic type structure, similarly to the reason why β-PbF_2 has a higher conductivity than BaF_2 [4]. This hypothesis has been corroborated by neutron diffraction of isotypic $PbSnF_4$ versus temperature, which has shown that the two fluorine sites in bridging positions get under populated when temperature increases, and the missing fluoride ions partially populate the spacing between tin layers, i.e. in the sheets of lone pair.

2 Experimental methods

2.1 Materials

The following reactants were used for the syntheses: SnF_2 99%, $BaCl_2{\cdot}2H_2O$ analytical grade, BaF_2 99% from Sigma–Aldrich, and doubly distilled or deionized water.

2.2 Precipitation reactions

Precipitation reactions were carried out either by adding a 1.274 M solution of $BaCl_2{\cdot}2H_2O$ to a saturated solution of SnF_2 (1.500 M) (Ba→Sn), or vice versa (Sn→Ba), on stirring, which resulted in the immediate precipitation of a white solid. The solid was filtered, washed with cold distilled water, and allowed to dry in air at ambient temperature. The

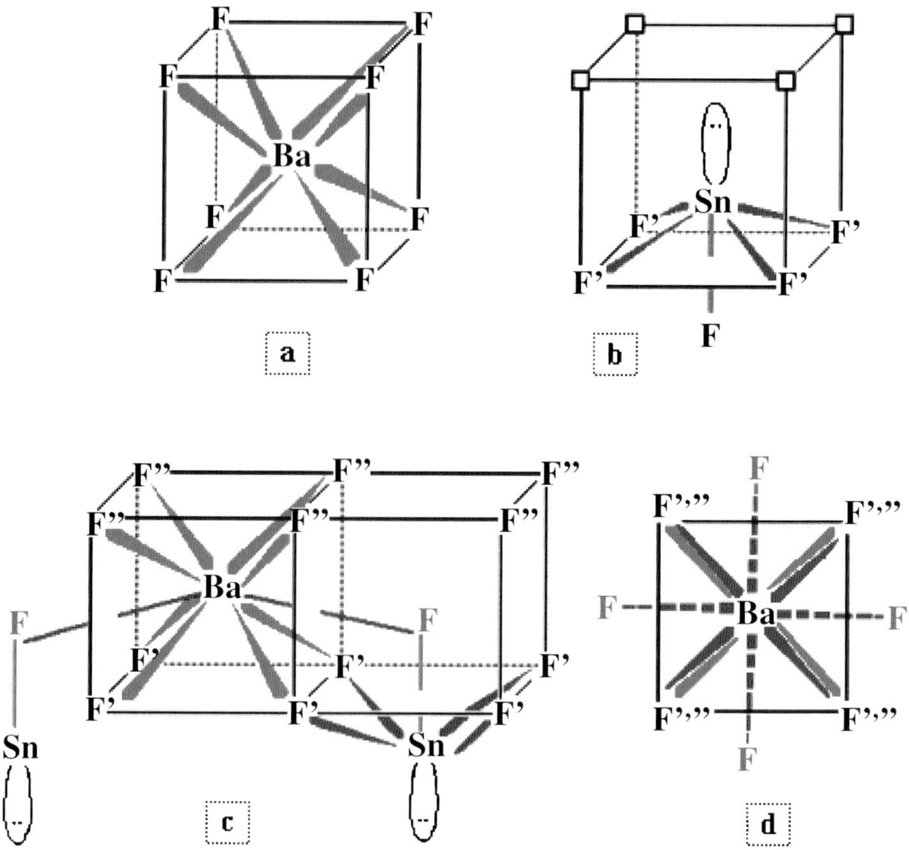

Fig. 2 Barium and tin coordination in BaF_2 and in $BaSnF_4$: **a** BaF_8 cube in BaF_2, **b** $SnFF'_4E$ pseudo-octahedron in $BaSnF_4$, **c** $BaF''_4F'_4F_4$ unit in $BaSnF_4$ and **d** *Top view* of the barium coordination in $BaSnF_4$ (*F'* and *F''*) are *superimposed*

dried solid (1.500 mg) was stirred in 60 ml of ultra pure water for a period of time and then washed with cold distilled water, and allowed to dry in air at ambient temperature.

2.3 Direct reactions at high temperatures

The starting materials used in these reactions are BaF_2 and SnF_2. The reagents were used without further purification since the X-ray diffraction shows no impurity line. The mass of each reactant used was determined from the reaction stoichiometery and desired mass of product. The reactants were placed in a copper tube under nitrogen, heated at 500°C for 4 h and cooled at ambient temperature, according to the method of Dénès [8].

2.4 Characterization

All samples were characterized by X-ray diffraction, on a Philips PW-1050 focusing powder diffractometer, using the K_α line of copper (λ=1.54178 Å). Mössbauer spectra were recorded on selected samples, using a nominally 20 mCi $Ca^{119m}SnO_3$ γ-ray source from Ritverc. The detector was a (Tl)NaI scintillation counter. The Doppler velocity (±10 mm/s) was generated by an Elscint driving system. The amplifier and the single channel analyzer are built in the Tracor Northern TN7200 multichannel analyzer, where the data were accumulated. Low temperature spectra were obtained in an APD Cryogenics helium closed cycle refrigerator equipped with a Displex® Mössbauer cold head. The GMFP software was used for fitting the spectra [9]. Isomer shits are referenced relative to the line of a $CaSnO_3$ standard absorber at ambient temperature.

3 Results and discussion

The source of barium was $BaCl_2{\cdot}2H_2O$, and the $BaCl_2/(BaCl_2{+}SnF_2)$ molar ratio $X = Ba/(Ba{+}Sn)$ was varied from 0.04 to 0.95 by adjusting the amount of $BaCl_2{\cdot}2H_2O$ solution used with a fixed amount of SnF_2 solution. The materials precipitated were studied mainly by X-ray powder diffraction. Phase diagram obtained by precipitation versus the stoichiometry X of the reaction mixture, and versus the method of addition is shown in Fig. 3.

The same materials are not always produced by the two methods of addition at the same given overall stoichiometry X of the reaction mixture. This can be accounted for by the fact that the local stoichiometry at the reaction site is usually not the same as the overall stoichiometry of the reaction mixture, and it is strongly dependent on the method of addition. The ions in excess at the reaction site can determine which product is formed at a given time in the precipitation reactions if stirring is unable to homogenize the reaction mixture before precipitation starts, regardless of the overall stoichiometry of the reactants, unless all the reactants in the receiving solution have been used up and no further reaction takes place. Other reactions parameters also have a strong influence on the materials that precipitate. The rate of addition of one reactant to the other must be chosen to be slow enough in such a way that the reactants being added have enough time to be mixed with the reactants of the receiving solution to form a fairly homogeneous mixture at the reaction site before much of them have reacted. Build up of the reactants being added occurs if a fast addition rate is used. A sufficiently slow rate of addition allows the reaction to take place smoothly with less chance to have multiphase products. In the present work, the rate of addition was set at 1.0 ml/min. for all reactions.

All three $MSnF_4$ compounds (M=Sr, Ba, and Pb) have been prepared by direct reaction between MF_2 and SnF_2 under dry conditions at high temperature. However, in 1975 all attempts by Dénès et al. [3] to obtain $BaSnF_4$ by precipitation reaction failed, and $PbSnF_4$ was the only $MSnF_4$ obtained by the wet method before the present work. For some precipitates obtained at high X values, the Search–Match gives $BaSnF_4$ as one possible phase present in the solid. No other match was satisfactory. When the new material discovered by Dénès et al. [13] previously, $BaSnClF_3{\cdot}0.8H_2O$, precipitates, no match was found to be reasonable except $BaSnF_4$ in some samples. The finding of $BaSnF_4$ at those stoichiometries very rich in Ba and Cl is rather surprising. It should be pointed out that, in all cases, $BaSnF_4$ was never obtained pure just by precipitation before stirring it in water (see Fig. 4). Furthermore, stirring the product formed by addition of a $BaCl_2$ solution to a SnF_2 solution, or vice versa, in the mother liquid for longer times does not give pure $BaSnF_4$. This is probably due to the fact that the presence of extra unused ions in the mother liquid, especially a large excess of chloride ions, favor the chloride–fluorides over chloride-free $BaSnF_4$. In an attempt to isolate $BaSnF_4$ from other phases, (X=0.90, Sn→Ba) was stirred in 60 ml water at room temperature for 46 h, since $BaSnF_4$ is poorly soluble in water. After filtration, the solid residue was washed with cold water and allowed to dry in air at ambient conditions. The result of the X-ray diffraction showed the solid residual was pure $BaSnF_4$ (Fig. 5) and a mass loss of 49.5% had taken place. All other phases had

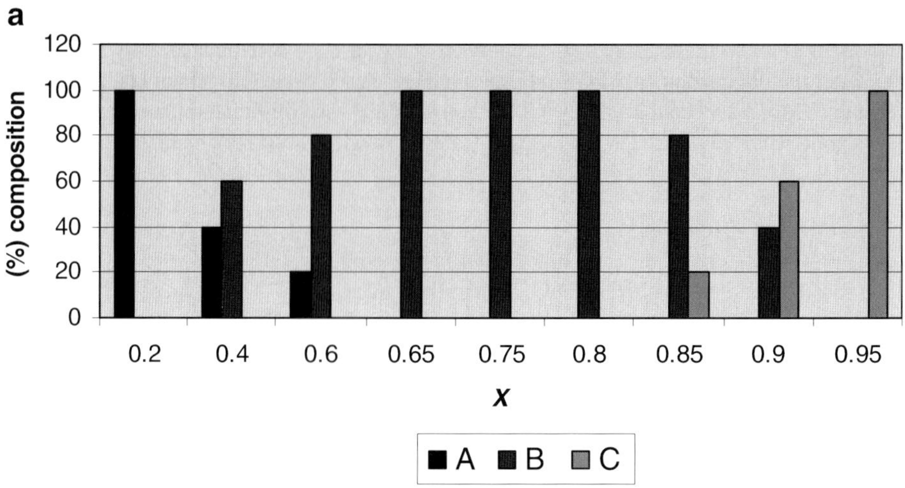

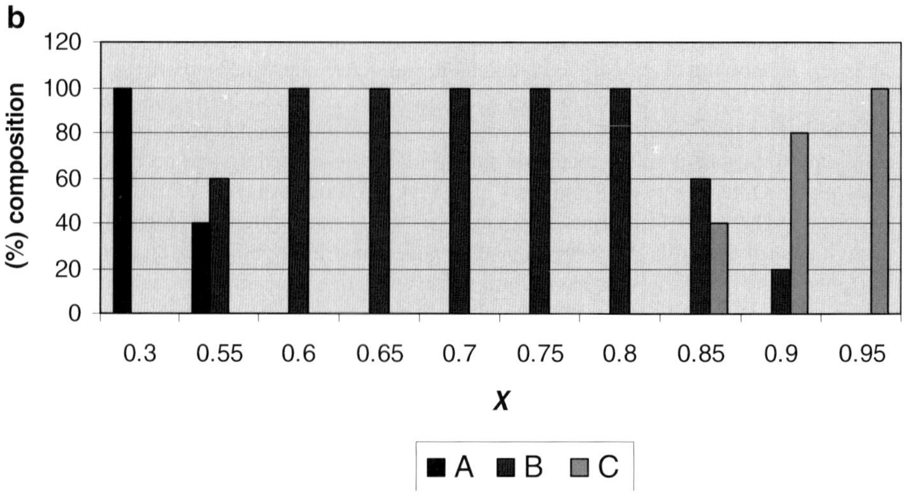

Fig. 3 **a** Estimated percentage of the phases obtained by precipitation reaction (Sn → Ba) versus *X*. Stirring was stopped at the end of the addition of the reactant, without additional water being added. *A*: $BaSn_2F_6$, *B*: $BaSnClF_3.0.8H_2O$, *C*: $Ba_{1-x}Sn_xCl_{1+y}F_{1-y}$. **b** Estimated percentage of the phases obtained by precipitation reaction (Ba→Sn) versus *X*. Stirring was stopped at the end of the addition of the reactant, without additional water being added. *A*: $BaSn_2F_6$, *B*: $BaSnClF_3.0.8H_2O$, *C*: $Ba_{1-x}Sn_xCl_{1+y}F_{1-y}$

disappeared. The ideal conditions of the preparations first established were found to be at room temperature, 1.200–1.500 g of chloride fluoride in 60 ml of water, and stirred for at least of 12 h.

The actual yield of $BaSnF_4$ obtained after stirring is plotted versus *X* on Fig. 6. It can be seen that the values vary from ca. 40 to 65%. Two possible mechanisms could be proposed to explain the formation of $BaSnF_4$ upon stirring in water: (1) the presence of soluble phases which could be dissolved during stirring, leading to a lower yield and disappearance of these phases from the X-ray diagram, and (2) the precipitated barium tin chloride fluorides product may be unstable in water, and therefore some species are leached out, leaving a product of different composition, namely pure $BaSnF_4$ or a mixture of

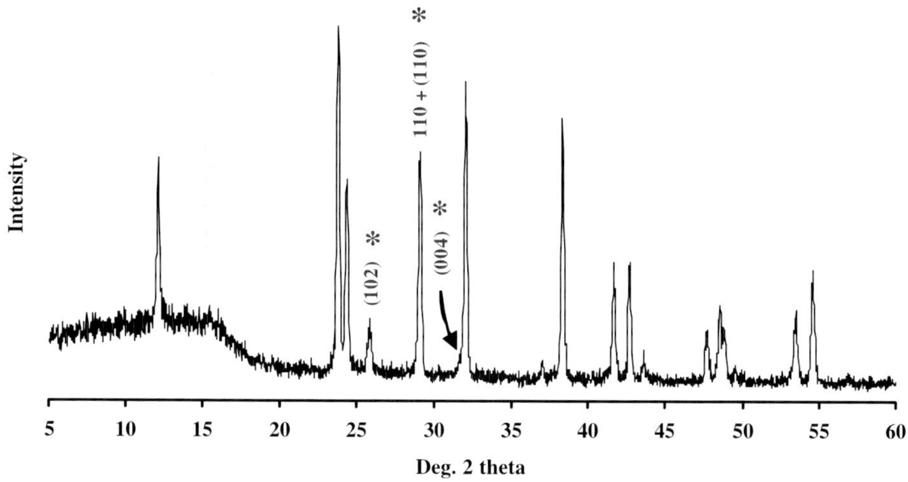

Fig. 4 X-ray diffraction pattern of a sample that contains two phases (X=0.95, Ba→Sn). The starred peaks correspond to $BaSnF_4$ that is present in traces amount, and the remaining peaks correspond to $Ba_{1-x}Sn_xCl_{1+y}F_{1-y}$. The (110) peak of $BaSnF_4$ and the (110) peak of $Ba_{1-x}Sn_xCl_{1+y}F_{1-y}$ overlap near perfectly well

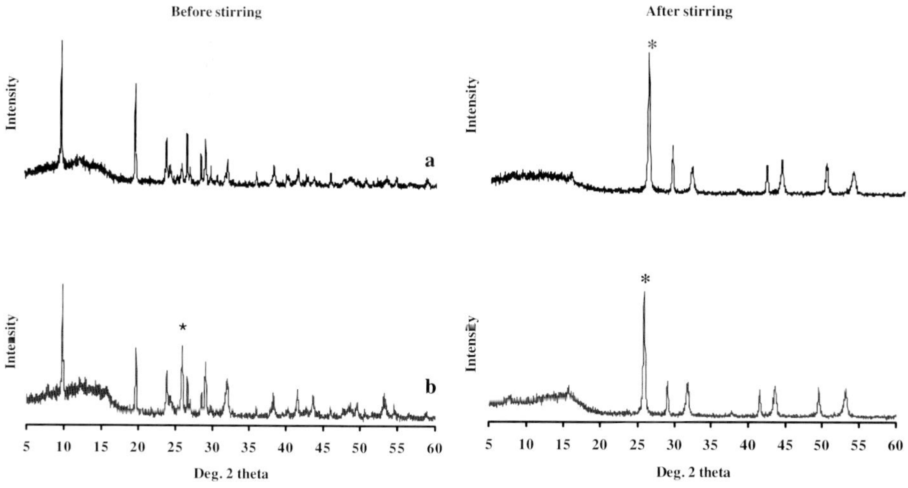

Fig. 5 X-ray diffraction patterns of the materials prepared by precipitation reaction at room temperature, **a** AM-1098 (X=0.835, Ba→Sn) and **b** AM-1103 (X=0.860, Ba→Sn) before (*left patterns*) and after stirring (*right patterns*) in water for 13 h (*: strongest peak of $BaSnF_4$ (102)). Before stirring, the precipitate is either **a** pure $BaSnClF_3.0.8H_2O$ or $BaSnClF_3.0.8H_2O$ mixed with some $BaSnF_4$

$BaSnF_4$ and $BaSn_2F_6$. This would also result in a lower yield. Process (1), i.e. the full dissolution of soluble products, is unlikely since the material is precipitated from aqueous medium, and therefore the soluble species would not have precipitated in the first place, although they might be more soluble in pure water than in the solution obtained after precipitation. In addition, if just the initial barium tin (II) chloride fluoride dissolved on stirring, the residue, i.e. $BaSnF_4$ and/or $BaSn_2F_6$ would be already present in the

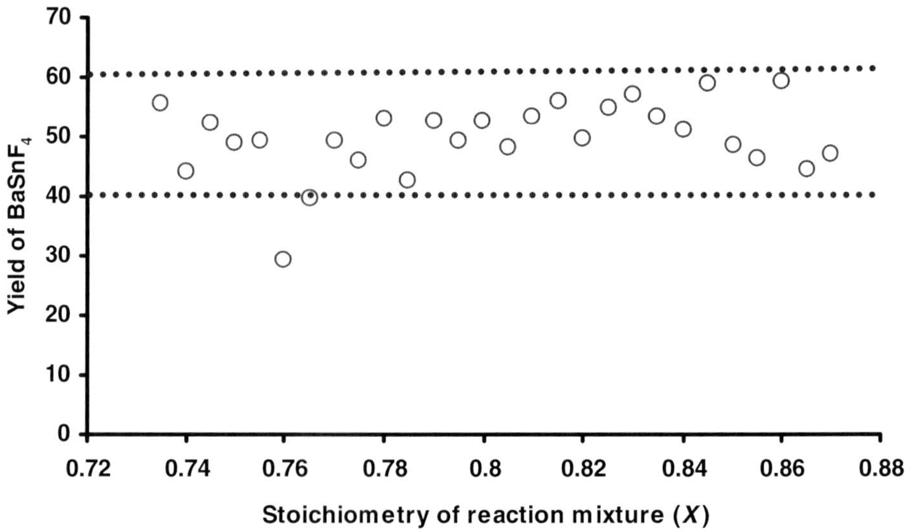

Fig. 6 Yield (%) of $BaSnF_4$ obtained from leaching, upon stirring $BaSnClF_3.0.8H_2O$ in water, versus the molar ratio *X* of $BaCl_2$ in the $SnF_2/BaCl_2$ reaction mixture used to precipitate $BaSnClF_3.0.8H_2O$

precipitated solid. However, each crystalline phase has its own set of diffraction lines, and therefore large amounts of $BaSnF_4$ and/or $BaSn_2F_6$ would be observed in the precipitates. Since when $BaSnF_4$ is present in the precipitates, it is in a minor amount, and many precipitates contain no $BaSnF_4$, and in addition, no $BaSn_2F_6$ is observed in the precipitates. Therefore process (2) is the most likely mechanism, i.e. the initial precipitate of barium tin (II) chloride fluoride undergoes a rearrangement in water, whereby all the chlorine and a large amount of barium are leached out of the solid and go in solution, resulting in a residue of $BaSn_2F_6$ or $BaSnF_4$, according to reaction 1 to 3.

$$2BaSnClF_3 \cdot 0.8H_2O(s) \xrightarrow{(\text{in } H_2O)} BaSnF_4(s) + Ba^{2+}(aq) + Sn^{2+}(aq) + 2Cl^-(aq) + 2F^-(aq) + 0.8H_2O(l) \quad (1)$$

$$2BaSnClF_3 \cdot 0.8H_2O(s) \xrightarrow{(\text{in } H_2O)} BaSn_2F_6(s) + Ba^{2+}(aq) + 2Cl^-(aq) + 0.8H_2O(l) \quad (2)$$

$$\frac{1}{x} Ba_{1-x}Sn_xCl_{1+y}F_{1-y}(s) \xrightarrow{(\text{in } H_2O)} BaSnF_4(s) + \frac{1-2x}{x} Ba^{2+}(aq) + \frac{1+y}{x} Cl^-(aq) + \frac{1-y-4x}{x} + F^-(aq) \quad (3)$$

The theoretical yield for reactions 1 to 3 is calculated to be 46, 67 and 26%, respectively for the Fluoride residue $BaSnF_4$, $BaSn_2F_6$ and $BaSnF_4$, respectively. However one must account for the possible partial solubility of $BaSnF_4$ and $BaSn_2F_6$, therefore actual yields must be expected to be lower than the theoretical values. Mixtures of $BaSn_2F_6$ and $BaSnF_4$ should give a theoretical yield between 46 and 67%, depending on the relative amounts of the two fluorides and that is in agreement with the results.

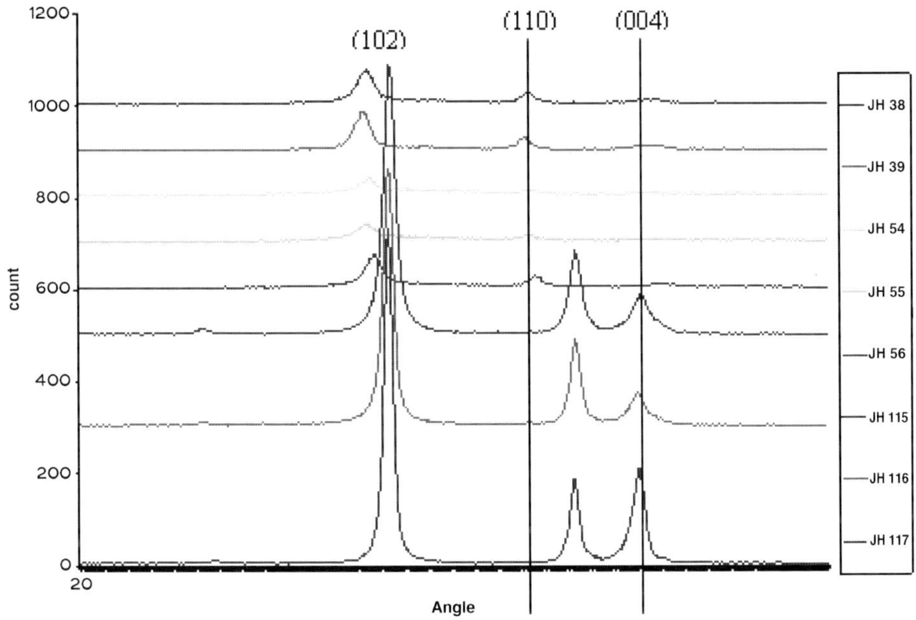

Fig. 7 Enlargement (20–35°) of the X-ray diffraction pattern of $BaSnF_4$ prepared by leaching reaction

Like the $BaSnF_4$ found mixed with the chloride fluoride precipitates, the $BaSnF_4$ obtained by stirring is obtained sometimes with narrow lines (large particle size), sometimes with broad lines (nanocrystalline), without a trend and with no apparent reason for the wide variability of particle dimension. Figure 7 compare the diffraction pattern of $BaSnF_4$ obtained by the dry method with one of the stirred samples that gives the narrowest lines (X=0.75, Ba→Sn, stirred for 24 h). It is clear that the linewidth of (110) is about the same on both, however, (004) is broader for the stirred sample. It can be concluded that heating BaF_2 and SnF_2 together at 500°C for 4 h results in a faster growth parallel to $\vec{c}$ than stirring $BaSnClF_3 \cdot 0.8H_2O$ in water for 24 h. The average particle dimensions parallel to $\vec{c}$ and perpendicularly to it, and the crystallographic constants, are given in Table 1 together with the unit-cell parameters versus the method of preparation. The unit-cell parameters of samples prepared by the various methods are close to one another. A significantly higher c parameter as can be seen by the shift of (004) to a lower angle in some samples where BaSnF4 was obtained after leaching for a period longer than 84 h cannot be rationalized at the present time, however, since these samples have the smallest particle size, it can be postulated that the particle size and the length of the periodicity parallel to $\vec{c}$ are related. This could be due to the shift of some fluoride ions other than the fluorine ion axially coordinating tin, for a fraction of tin, moving in the interlayer spacing where the layers of lone pairs are located. This is in agreement with the well known fact that the presence of large number of defects results in an increase of unit-cell size. This would create locally a situation similar to the fluoride position in BaF_2. Such a partial transfer has been shown by neutron diffraction to take place in $PbSnF_4$, and has been held responsible for its exceptionally high conductivity. The presence of this phenomenon would be in agreement with the cubic $\mu\gamma$-$BaSnF_4$ obtained by ball-milling. The sample prepared by solid state reaction has also a slightly larger c parameter. Figure 7 show that the a parameter is about the same for the three samples, whereas c varies. This

Table 1 Unit-cell constants and particle dimension anisotropy of $BaSnF_4$ versus the method of preparation

Sample ID	Method of preparation	Stirring period (h)	a (nm)	c (nm)	V (nm^3)	$c/2a$	$d\|$ (nm)	$d\perp$ (nm)	$d\|/d\perp$	Δd (nm) * 10^2
Dénès et al. [3]	Direct reaction	–	0.4356	1.1289	0.214	1.30	–	–	–	2.60
Muntasar [14]	Direct reaction	–	0.4287	1.1336	0.208	1.32	46.50	46.20	1.1	2.38
JH120	Direct reaction	–	0.4228	1.1480	0.205	1.36	34.93	27.07	1.290	2.51
JH38	Leaching reaction Ba→Sn	72	0.4342	1.1360	0.214	1.31	11.32	28.80	0.393	2.30
JH39	Leaching reaction Ba→Sn	72	0.4370	1.1360	0.217	1.30	10.42	26.19	0.398	2.50
JH54	Leaching reaction Ba→Sn	84	0.4342	1.1320	0.213	1.30	6.21	21.49	0.289	2.40
JH55	Leaching reaction Ba→Sn	84	0.4370	1.1360	0.217	1.30	7.47	15.56	0.480	2.50
JH56	Leaching reaction Ba→Sn	84	0.4327	1.1320	0.212	1.31	13.44	28.20	0.477	2.30
JH115	Leaching reaction Sn→Ba	24	0.4214	1.1440	0.203	1.36	22.68	29.49	0.769	1.20
JH116	Leaching reaction Sn→Ba	24	0.4228	1.1440	0.205	1.35	25.83	35.08	0.736	1.39
JH117	Leaching reaction Sn→Ba	24	0.4228	1.1480	0.205	1.36	28.95	38.38	0.754	1.29
Average direct reaction		–	0.4292	1.138	0.210	1.33	34.93	27.07	1.290	2.50
Average leaching reaction Ba→Sn		–	0.4350	1.134	0.215	1.30	9.77	24.05	0.407	2.40
Average leaching reaction Sn→Ba		–	0.4223	1.145	0.204	1.36	25.82	34.32	0.753	1.29

is not too surprising. Long distance polymerization parallel to ($\vec{a}$, $\vec{b}$) keeps the size of the ($\vec{a}$, $\vec{b}$) plane constant, since it is mainly determined by the relative sizes of the Ba^{2+} and F^- ions, whereas the loose interaction between the sheets parallel to ($\vec{a}$, $\vec{b}$), allows for easy variations of the inter-sheet spacing, and thereby of the c parameter.

Particle size anisotropy is always observed in $BaSnF_4$, whereby .the particle size is smaller in the direction parallel to the $\vec{c}$ axis than in the perpendicular direction. This is explained by the layered structure of the material, due to the sheets of tin (II) lone pairs perpendicular to $\vec{c}$ that act as perfect cleavage planes. The interlayer spacing was found to vary from one preparation to another due to weak interlayer directions, whereas the distances in the perpendicular direction are stable. Wide variation of particle sizes of $BaSnF_4$ is observed, while $BaSn_2F_6$ has larger crystallites.

These erratic occurrences of $BaSnF_4$ particle size lack of trend is undoubtedly due to "hidden" reaction and stirring parameters that have not been identified yet. The reaction that takes place consists of a large part of the Ba and all the Cl contained in the sample being leached out, and a 40 to 60% mass loss is observed. Breaking up M–Cl bonds (M=Ba and Sn) preferentially to M–F is understood in terms of bond strength, where the M–Cl bonds being weaker. $BaSnF_4$ has been identified as having unusually high temperature dependence of the quadrupole splitting (Table 2). This has been explained by its very highly anisotropic thermal expansion and the extraordinary high temperature dependence of

Table 2 Mössbauer parameters of $BaSnF_4$ versus the method of preparation (*D* High temperature dry method, *S* Chloride fluoride stirred in filtrate, *L* Chloride fluoride leached by stirring in water)

AM sample number	Preparation method	δ (mm/s)	Δ (mm/s)	Sn(IV) signal (%)	*T* (K)
456	D	3.289	1.530	14.24	298
10991321	S	3.289	1.530	30.280	298
108918H	S	3.359	1.530	28.740	298
$BaSnF_4$	L	3.450	1.953	4.500	76
α-$PbSnF_4$	L	3.24	1.560	–	298
α-SnF_2	L	3.430	1.532	1–2	298

the thermal expansion coefficient, and this is accounted for by the unconventional distortion of the tin site, brought about by the large size of the Ba^{2+} ion [7, 11].

In cases where the non-bonded electron pair on Sn(II) is stereochemically inactive and would contribute to the conduction band, the material might be considered to be a strong electronic conductor. On the other hand, if the two electrons are in a localized orbital with considerable *p* character, the nature of the conductivity is quite different. These possibilities are easily examined by ^{119}Sn Mössbauer spectroscopy, since this technique provides a convenient way of probing the stereochemical activity of the tin lone pair. The $BaSnF_4$ prepared by both methods, i.e. precipitation followed by stirring, and by high temperature solid state reaction, were studied by Mössbauer spectroscopy. The Mössbauer spectrum of $BaSnF_4$ prepared by both methods have a quadrupole doublet characteristic of divalent tin in a non-cubic environment (Fig. 8). The relevant Mössbauer parameters are given in Table 2. The small peak at around zero velocity is a Sn(IV) impurity. This impurity, is also present in the starting SnF_2 but it is not detected by diffraction either because of the small amount present and/or its possible amorphous nature [10, 12].

The isomer shift and quadrupole splitting for $BaSnF_4$ prepared here by the two methods (direct reaction and stirring) are very similar to one another, and show that there is no major difference in the local tin situation (lone pair stereoactivity, bonding) between the two samples. They are also very similar to the literature values for α-$PbSnF_4$ [12]. This is not surprising, since the two phases are isostructural. The quadrupole splitting is the same as for SnF_2, however the isomer shift is lower [10]. This is also not surprising since the tin coordination is not the same in the two structures. The lower isomer shift of $MSnF_4$ shows a lower *s* electron density in the lone pair, and therefore more in the bonding pairs. This should result in a higher *p* and *d* orbital contribution to the lone pair, hence a higher quadrupole splitting for $MSnF_4$. However, the quadrupole splitting of $BaSnF_4$ recorded in this work and that of α-SnF_2 are the same, and that of α-$PbSnF_4$ is also nearly the same. This can be accounted for by the different structures of $MSnF_4$ and α-SnF_2, hence a different $(V_{zz})_{latt.}$ that, when added to their different $(V_{zz})_{val.}$, give the same total V_{zz}. Such cancellation of the effects of the difference of $(V_{zz})_{val.}$ and $(V_{zz})_{latt.}$ is not uncommon, and has also been observed by Dénès to take place between the two tin sites of α-SnF_2 [10]. The isomer shift and quadrupole splitting of $BaSnF_4$ reported in the literature are significantly higher, especially the quadrupole splitting [5, 6]. This can be accounted for by the sample temperature during data collection (80 K) whereas all the others were measured at ambient conditions. The higher isomer shift at 80 K is due to the *second order Doppler shift*, i.e. the contribution of the mean square velocity of the Mössbauer nuclide to the isomer shift. This contribution is negative, and increases with increasing temperature, i.e. an increase of temperature results in a decrease of isomer shift. A decrease of quadrupole splitting with increasing temperature has also been observed on α-SnF_2, however, the effect

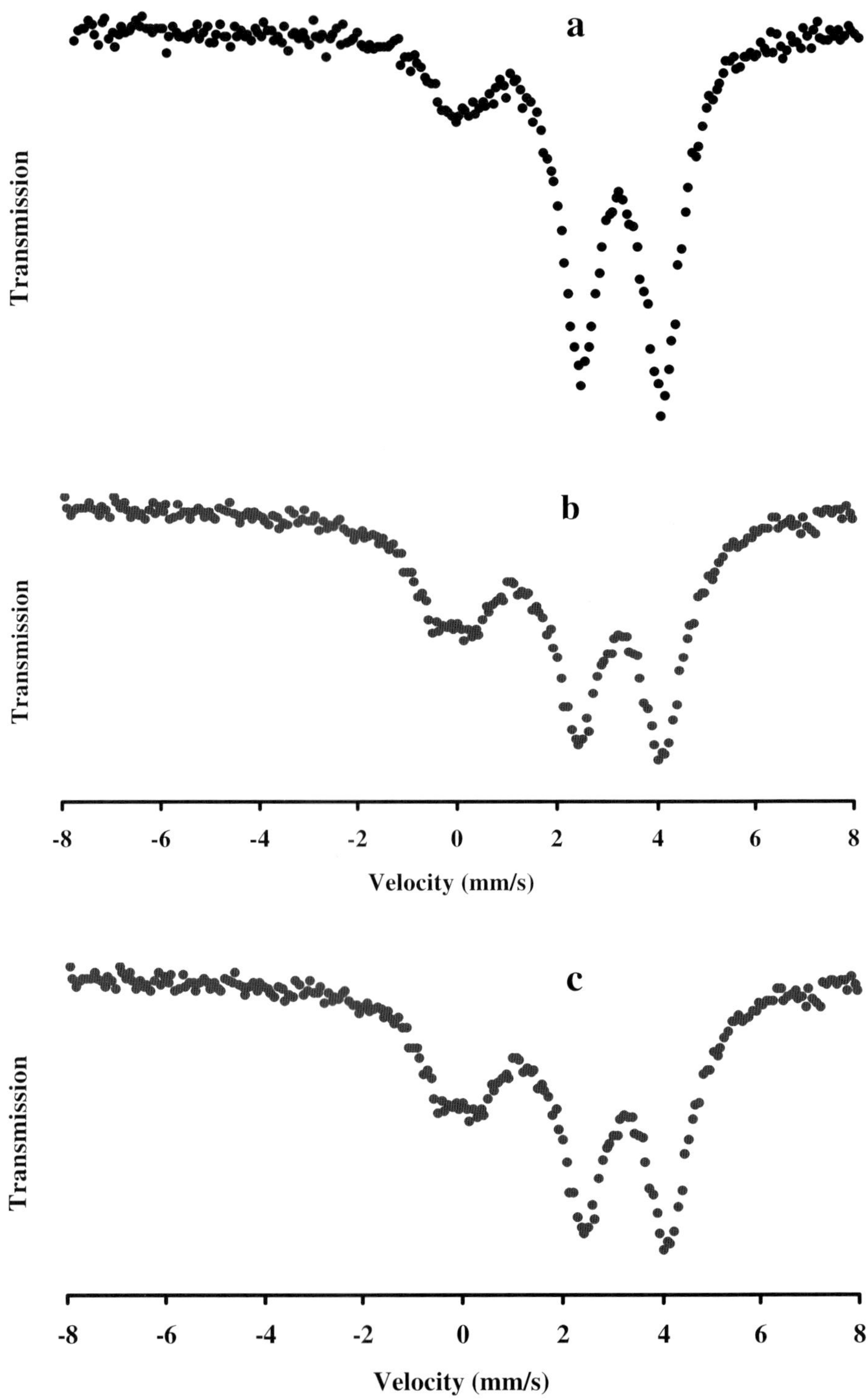
a
b
c
Transmission
Transmission
Transmission
-8 -6 -4 -2 0 2 4 6 8
Velocity (mm/s)
-8 -6 -4 -2 0 2 4 6 8
Velocity (mm/s)

Fig. 8 Tin-119 Mössbauer spectrum of $BaSnF_4$, **a** prepared by direct reaction at 500°C for 2 h, **b** and **c** prepared by precipitation (X=0.800 and 0.845, Ba→Sn), respectively, followed by stirring in water for 22 h

is much lower than in $BaSnF_4$ [10]. The exceptionally large quadrupole splitting of $BaSnF_4$ at low temperature has been explained by the unusually highly flattened tin coordination relative to α-$PbSnF_4$, and this originates in the large size of the Ba^{2+} ion, which is also responsible for the large tetragonal distortion, i.e. the small $c/2a$ ratio [3, 11]. X-ray diffraction versus temperature has shown that the thermal expansion is very highly anisotropic, with $\alpha_3 > \alpha_1$, and becomes more and more anisotropic when temperature increases, where α_1, α_2 and α_3 are the thermal expansion coefficients parallel to $\vec{a}$, $\vec{b}$ and $\vec{c}$, respectively ($\alpha_1 = \alpha_2$ by symmetry, in tetragonal). It results that, as temperature increases, the tetragonal distortion decreases, i.e. $c/2a$ increases, the site distortion at tin decreases (less flattened polyhedron), and therefore the quadrupole splitting should decrease, which is in agreement with our observations. The unusually high thermal expansion anisotropy results in an exceptionally high decrease of quadrupole splitting with increasing temperature. The large isomer shift and quadrupole splitting of $BaSnF_4$, α-$PbSnF_4$ and α-SnF_2 are in agreement with the presence of divalent tin, the lone pair of which is highly stereoactive and dominates the value of $(V_{zz})_{val.}$ The high stereoactivity of the lone pair prohibits the $5s^2$ electrons from being mobile, and is in agreement with the ionic nature (F^-) of the charge transport in both $MSnF_4$ and SnF_2 and their low transport number for electrons ($\tau_e \leq 0.01$).

4 Conclusion

It has been shown that $BaSnClF_3 \cdot 0.8H_2O$ is always fully decomposed when stirred in water, leaving $BaSnF_4$ pure phase. Particle size anisotropy is always observed in $BaSnF_4$, whereby the particle size is smaller in the direction parallel to the $\vec{c}$ axis than in the perpendicular direction. This is explained by the layered structure of the material, due to the sheets of tin (II) lone pairs perpendicular to $\vec{c}$ that act as perfect cleavage planes. The interlayer spacing was found to vary from one preparation to another due to weak interlayer directions, whereas the distances in the perpendicular direction are stable. Wide variation of particle sizes of $BaSnF_4$ is observed, while $BaSn_2F_6$ has larger crystallites.

At this point, many occurrences seem to be erratic, such as the presence of $BaSnF_4$ particle size. This lack of trend is undoubtedly due to "hidden" reaction and stirring parameters that have not been identified yet. The reaction that takes place consists of a large part of the Ba and all the Cl contained in the sample being leached out, and a 40 to 60% mass loss is observed. Breaking up M–Cl bonds (M=Ba and Sn) preferentially to M–F is understood in terms of bond strength, where the M–Cl bonds being weaker.

$BaSnF_4$ has been identified as having unusually high temperature dependence of the quadrupole splitting. This has been explained by its very highly anisotropic thermal expansion and the extraordinary high temperature dependence of the thermal expansion coefficient, and this is accounted for by the unconventional distortion of the tin site, brought about by the large size of the Ba^{2+} ion.

Acknowledgements This work was made possible by the support of Concordia University, the Natural Science and Engineering Research Council of Canada, and Charles River Laboratories. Grateful thanks are also due to the Procter and Gamble Co. (Mason, Ohio) for supporting our Mössbauer laboratory.

References

1. Donaldson, J.D., Senior, B.J.: J. Chem. Soc. **A**, 1821 (1967)
2. Dénès, G., Muntasar, A., Zhu, Z.: Hyperfine Interact. C **1**, 468 (1996)
3. Dénès, G., Pannetier, J., Lucas, J.: C. R. Acad. Paris C **280**, 831 (1975)
4. Dénès, G.: Solid State Ionics IV, Mater. Res. Soc. Proc. **369**, 295 (1985)
5. Birchall, T., Dénès, G., Ruebenbauer, K., Pannetier, J.: Hyperfine Interact. **30**, 167 (1986)
6. Dénès, G., Yu, Y.H., Tyliszczak, T., Hitchcock, A.P.: J. Solid State Chem. **91**, 1 (1991)
7. Dénès, G., Birchall, T., Sayer, M., Bell, M.F.: Solid State Ion. **13**, 213 (1984)
8. Denes, G.: J. Solid State Chem. **77**, 54 (1988)
9. Ruebenbauer, K., Birchall, T.: Hyperfine Interact. **7**, 125 (1979)
10. Birchall, T., Dénès, G., Ruebenbauer, K., Pannetier, J.: J.C.S. Dalton, 1831 (1981)
11. Shannon, R.D., Prewitt, C.T.: Acta Crystallogr., B **25**, 725 (1969)
12. Birchall, T., Dénès, G., Ruebenbauer, K., Pannetier, J.: J.C.S. Dalton, 2296 (1981)
13. Dénès, G., Muntasar, A.: Hyperfine Interact. **153**, 91 (2004)
14. Muntasar, A.: PhD thesis, pp. 158–192. Concordia University, Montreal, Quebec, Canada (2002)

Hyperfine Interact (2006) 170:159–167
DOI 10.1007/s10751-006-9462-7

The Mössbauer community in the USA

J. G. Stevens · A. M. Khasanov · N. F. Hall ·
I. A. Khasanova

Published online: 9 January 2007

Abstract Scientists in the United States assumed major roles in developing the Mössbauer community during its early years. However, since the termination of the Mössbauer Effect Methodology meetings in 1976, there has been little in the way of regular Mössbauer meetings in the United States. Nevertheless, there is an active United States Mössbauer community, as noted by the number of annual publications – 156 in 2004. In recent decades, attendance of Mössbauer researchers from the United States at the International Conferences on the Applications of the Mössbauer Effect (ICAME) has been far below what would be expected from the number of contributions in the Mössbauer literature. Attempts have been made, unsuccessfully, to arrange for regular Mössbauer meetings. Models for possible future Mössbauer meetings of US scientists are discussed, including a regular biannual meeting, and another being a virtual Mössbauer conference. Also discussed are other models to maintaining an active Mössbauer community in the United States, making use of information technologies that are available to us along with other resources we can use.

Key words Mössbauer · analysis

1 History of Mössbauer spectroscopy in the United States

The history of Mössbauer effect research in the United States necessarily begins with Rudolf Mössbauer, who first observed the effect in 1957 while still a graduate student at the Technical Academy of Munich. He received his PhD in 1958, and came to the California Institute of Technology (CalTech) as a Research Fellow in 1960. In 1961 he was awarded the Nobel Prize for Physics, jointly with Professor Hofstadter, at which time he was also appointed Professor of Physics at CalTech. Professor Mössbauer returned to Munich a few

J. G. Stevens (✉) · A. M. Khasanov · N. F. Hall · I. A. Khasanova
Mössbauer Effect Data Center, University of North Carolina at Asheville,
One University Heights, CPO 2311, Asheville, NC 28804-8511, USA
e-mail: stevens@unca.edu

years later to join the Physics Faculty at the Technical Academy, but briefly returned to CalTech as a Visiting Professor in 1964.

The vigor of the early US Mössbauer community can be viewed through the lens of the *Mössbauer Effect Data Index* – an ambitious project undertaken by US researchers Arthur Muir, Ken Ando, and Helen Coogan of the North American Aviation Science Center to compile an organized cumulative index to published research results relating to the Mössbauer effect. The first *Mössbauer Effect Data Index* was published in 1966, and covered the Mössbauer literature published from 1958 through 1965. Subsequently, the *Index* was published by the Mössbauer Effect Data Center in Asheville, North Carolina, with various editors through 1976, at which point the *Index* was replaced by the publication of the *Mössbauer Effect Reference and Data Journal (MERDJ)*. The *MERDJ* is published 10 times per year, with an annual Index, and continues through the present time.

At the forefront of US Mössbauer research in the early years was the Argonne National Laboratory in Illinois. Important Argonne researchers include Gopal Shenoy, Gil Perlow, Mike Kalvius, and Fritz Wagner. Stan Ruby, who sadly passed away in 2004, was a major figure in Mössbauer research at Argonne. Clive Wynter and Ercan Alp have dedicated this meeting to Stan Ruby's honor, and Gopal Shenoy has spoken on Stan's life and legacy. My first 3 years in academia were spent with Stan, and he has influenced my life and career in a profound way. Argonne continues to be a leader in US Mössbauer research today; two of the most productive US Mössbauer researchers today in terms of number of publications are Wolfgang Sturhahn and Ercan Alp, both from Argonne. Argonne is the second most productive US institution in terms of number of publications, having over 40 Mössbauer-related papers in the past 5 years.

The United States also hosted the first international Mössbauer scientific meeting. Described by Hans Frauenfelder in the proceedings as "an informal meeting on [the] Mössbauer effect" [1], the Allerton House Mössbauer meeting is now considered to be the first in the series of conferences and meetings that eventually evolved into the ICAME (International Conference on the Applications of the Mössbauer Effect) series. The meeting was held for 2 days (June 6 and 7, 1960) in the Allerton House at the University of Illinois. According to Frauenfelder, "Only five weeks elapsed between the original suggestion for holding such a meeting and the meeting itself" [1]. The meeting was attended by 89 researchers, most of whom were from the United States, but also attending were scientists from Canada, France, India, Israel, Sweden, and the United Kingdom.

Another long-running conference series held in the United States was the Mössbauer Effect Methodology symposia, which ran from 1965 to 1976. The symposia were all held in New York City, with the exception of two held in Chicago in 1968 and again in 1974. Dean Taylor, writing about the symposia series in an issue of the *MERDJ*, describes it as follows: "In the mid-60's Stan Ruby approached Irwin Gruverman of NENC, the New England Nuclear Corporation, suggesting that NENC sponsor a one-day Mössbauer Effect Methodology conference (MEM) in association with the winter meeting of the APS [American Physical Society].... The series began on January 26, 1965, in New York City with free registration for all 250 participants. The 15 papers were presented in an afternoon/evening format. As needed, NENC paid the expenses of the chosen speakers, who generally were from labs in the USA.... Manuscripts were published in *Mössbauer Effect Methodology Volume 1* (through Volume 10) by Plenum Press, New York, edited by I. Gruverman. The meeting was a huge success, and it was subsequently held annually until 1976 (except 1972 and 1975). Then the energy crisis of the 1970s, the sharp decline in funding for basic sciences, perhaps a perceived lack of newness in the discipline, the decline of attendance at MEM, and the renewal of the International Mössbauer series of conferences meant that MEM had served its purpose." [2].

One of the first books published on the Mössbauer effect was entitled "The Mössbauer Effect: A Review – with a Collection of Reprints," authored and edited by Hans Frauenfelder of the University of Illinois. Published in 1962, only 4 years after Professor Mössbauer's discovery, the book contained an introduction to the Mössbauer effect, sections covering theory, experimental apparatus and problems, nuclear properties, general physics, and solid-state physics, a bibliography, and a collection of reprints and translations of seminal Mössbauer research papers.

In 1993, the International Conference on the Applications of the Mössbauer Effect was held in North America for the first time since 1963, when the Third International Conference on the Mössbauer Effect was held in Ithaca, New York. ICAME '93 was held in Vancouver, Canada, on the campus of the University of British Columbia. This conference was attended by 349 international participants, of which 72 were from the United States.

Mössbauer research has been featured in focused sessions and symposia at American Physical Society (APS) meetings. For example, in 1991 there was a symposium on Application of Mössbauer Spectroscopy in Condensed Matter, and in 1994 a Focused Session on Mössbauer Spectroscopy. Such focused sessions and symposia have declined in recent years. However, Mössbauer results are still presented at APS meetings. At the March 2005 APS Meeting, seven Mössbauer papers were presented, and 15 were presented at the March 2004 APS Meeting.

Much like APS meetings, Mössbauer research has been well represented at National Meetings of the American Chemical Society (ACS). An early symposium on the Mössbauer Effect and its Application in Chemistry was held in 1966, a 1980 symposium focused on Recent Chemical Applications of Mössbauer Spectroscopy, and a symposium on Twenty-Five Years of Chemical Mössbauer Spectroscopy was presented at an ACS meeting in 1984. The ACS is also a major sponsor of the International Chemical Congress of Pacific Basin Societies, which hosted the first Symposium on the Industrial Applications of the Mössbauer Effect in 1984 in Honolulu, Hawaii. That symposium has continued in other venues and is now held every 4 years, known as the ISIAME (International Symposium on Industrial Applications of the Mössbauer Effect) series. The most recent ISIAME was held in Madrid in 2004. Unfortunately, the number of special Mössbauer symposia at ACS meetings has all but disappeared in recent years. However, Mossbauer results *are* being presented at ACS meetings. Twelve Mössbauer-related papers were presented at the 228th ACS National Meeting in August 2004, and seven each were presented at the 229th and 230th National Meetings held in 2005. Also, at least six Mössbauer papers were presented at the most recent Pacifichem conference held in Hawaii in December 2005.

The only continuing semi-regular US Mössbauer meeting in recent times that enables US Mössbauer researchers to continue their scientific dialogue has been the the Nassau Mössbauer meetings organized by Clive Wynter and Ercan Alp. The first Nassau Mössbauer Conference was held in 1976, and was organized by Clive Wynter and Rolfe Herber. Twenty-two papers were presented at this first Nassau meeting. The second conference was organized by Clive and Ercan Alp in 1993, and was attended by some 30 Mössbauer spectroscopists. The conference was revived 10 years later in 2003, again organized by Clive and Ercan, and was attended by about 35 scientists, and the most recent conference was held in January 2006, attended by 42 scientists.

The challenge facing US Mössbauer scientists is to find a way to regain cohesion for the US Mössbauer community. We need to ensure that we, as a community, can overcome

Table 1 Analysis of US Mössbauer researchers compared to all Mössbauer researchers

	United States	Worldwide	Percentage of US to world total
Very active Mössbauer researchers	28	364	7.7%
Active Mössbauer researchers	64	698	9.2%

isolation, create an ongoing dialogue, and more fully participate in the global Mössbauer and general scientific communities.

2 Analysis of US Mössbauer research

The United States remains an active national player in worldwide Mössbauer research. According to the records of MEDC, the total number of recent US scientists involved in Mössbauer research is 343. This number is based on scientists who have published at least two Mössbauer-related papers in the past 5 years. These scientists work at 141 institutions (US institutions that have published at least two papers in the past 5 years). Of those US Mössbauer scientists, 64 are what the Center terms "active researchers" and have published five or more papers in the past 5 years, and 28 are termed "very active researchers" and have published 10 or more papers in the past 5 years (Table 1). When compared with global statistics, we find that 7.7% of global "very active" researchers are from the US, and 9.2% of the global "active" researchers are from the US.

Seeing that US researchers are still active in Mössbauer research, we can examine how US publications compare to the rest of the world (Fig. 1). US researchers contribute from 130 to 220 Mössbauer publications annually; this represents between 13 and 17% of total Mössbauer publications worldwide. Note that Fig. 1a shows a decline in US Mössbauer publications beginning in 2002. In 2002, US researchers published 218 papers, which dropped to 137 in 2003 and 133 in 2004. Our records show only 73 papers from US researchers in 2005, but these numbers are somewhat skewed as we are still processing 2005 papers.

Now we can compare the US contribution of general Mössbauer research to US participation in the ICAME conferences. Figure 2 compares the number of US participants at each of the last six ICAME conferences to the number of active researchers in the US at that time. For instance, at the conference held in Rimini in 1995, 23 participants were from the US, while at the same time there were 124 active US researchers. We can see that over the past 10 years, only 19–53% of US active researchers participate in ICAME conferences. Note in Fig. 2 a decrease in the number of active US researchers beginning in 1995 (the Rimini conference). Also note that there were 72 US participants at the Vancouver ICAME, held in 1993 (the conference immediately preceding the Rimini conference in 1995). The number of US participants dropped from 72 to 23 in the 2 years between the conferences. This, of course, is largely due to the logistics and costs of traveling abroad to attend a major Mössbauer conference, and a very good indication that interest may exist for continuation of some type of regular gathering of US Mössbauer researchers.

Figure 3 shows the percentage of total US participants at the last six ICAME conferences compared to the percentage of US contributions to publications in that year. We can see that US participation in ICAME conferences is two to three times lower than the

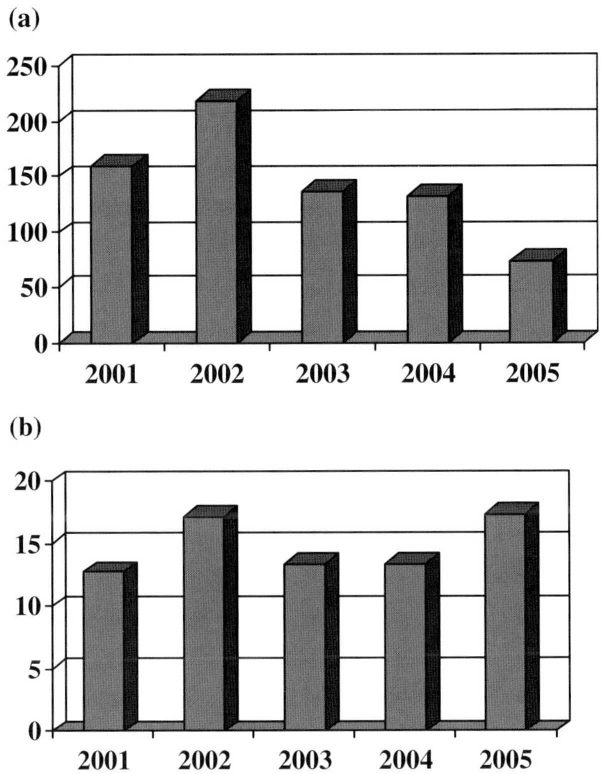

Fig. 1 US contribution to Mössbauer research for the period 2001–2005: **a** yearly number of Mössbauer-related publications by US researchers; **b** yearly percentage of worldwide Mössbauer publications attributable to US researchers

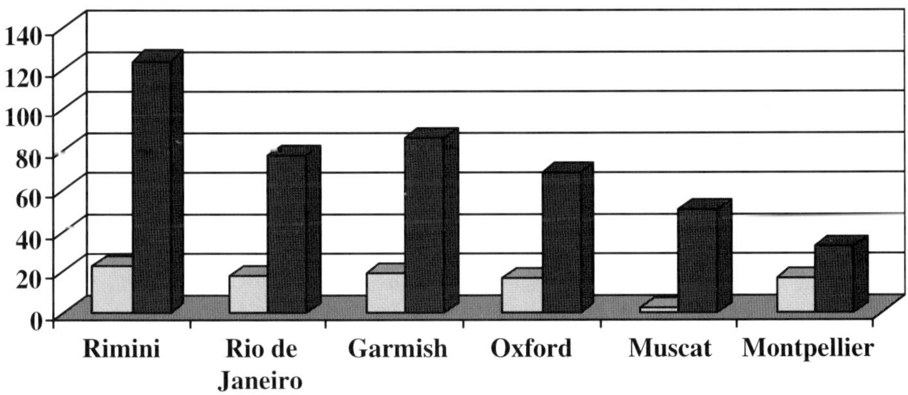

Fig. 2 US participation in the last six ICAME conferences (*light bar*) compared to number of active US researchers in that year (*dark bar*)

corresponding US contribution to publications in that year. This speaks to the relative isolation of US researchers in the Mössbauer field – while research continues, as indicated by the number of publications, many of the researchers are, for whatever reason, simply not participating in the major international Mössbauer conferences.

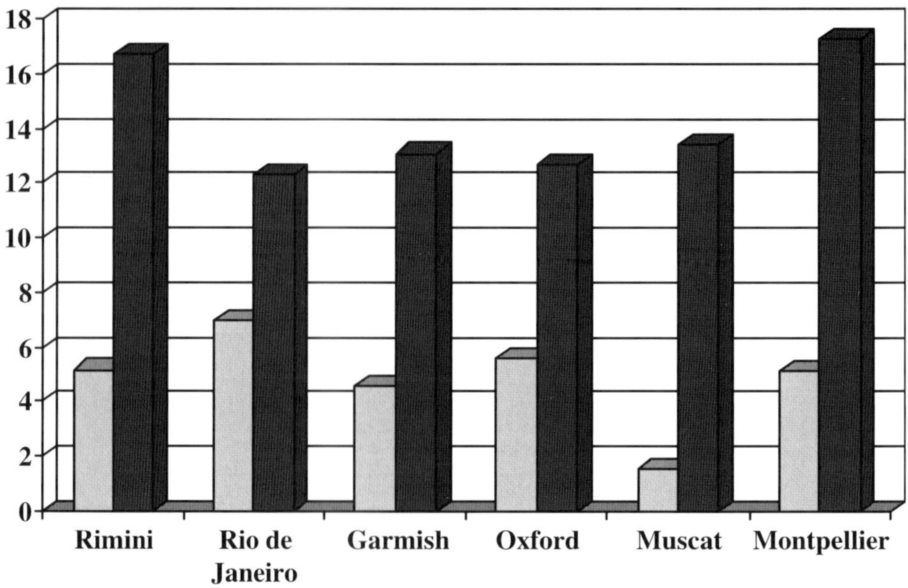

Fig. 3 Percentage of total US participants at the last six ICAME conferences (*light bar*) compared to the percentage of US contributions to worldwide publications in that year (*dark bar*)

3 Possibilities for strengthening the US Mössbauer community

The need for a strengthening of the US Mössbauer community to overcome isolation and begin a dialogue is apparent. But how can we accomplish this? One way may be to continue our face-to-face, traditional meetings, perhaps by continuing this Nassau conference series. Other suggestions may be to:

- Create an e-mail listserve for the US Mössbauer community that would allow subscribers to communicate with one another via e-mail in a timely manner
- Create a regional newsletter for the US Mössbauer community, one that could perhaps be published electronically on a quarterly or semiannual basis, to keep us all informed of each other's research and other interests.
- Another idea is the creation of a regional US Web Page. This Web Page could be served by the new Web site http://www.mossbauer.org created recently by MEDC and IBAME, and it would co-exist with other regional sites from the international Mössbauer community.
- Finally, the idea of a virtual conference has been discussed.

Virtual Conference First, we can review some possible models for the virtual conference idea that has been circulating in the community (Fig. 4). The first model, Model A, would involve presentation via a conference Web site, which would then be discussed by participants via a discussion forum on the same Web site. In this model, a Web server would be utilized whereby a presenter would prepare his or her presentation and upload it to the server. The other participants would access the presentation from the Web server via a standard Web browser on their computers. Discussion of the presentation would be via forum on the same Web page. Model B differs only in the way a discussion is maintained.

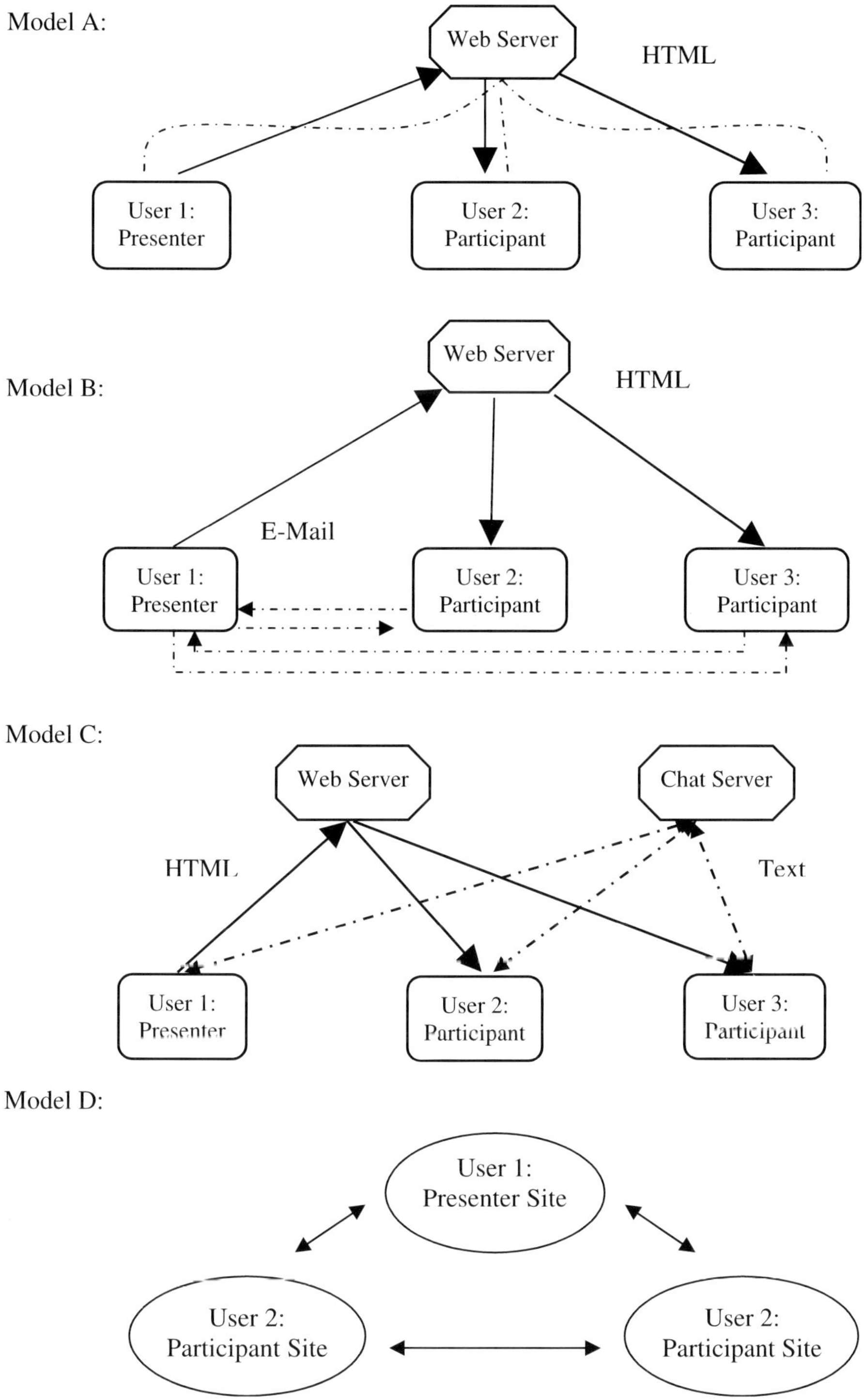

Fig. 4 Models for a possible virtual conference

Table 2 Comparison of the virtual conference models

	Price	Duration	Number of participants	Lag time
Model N	\$500–\$2,000	8 h/5 days	20–200	0
Model A	\$0–\$50	24 h/ 365 days	Unlimited	<3 h
Model B	\$0–\$50	24 h/ 365 days	Unlimited	<1 h
Model C	\$0–\$50	24 h/ 365 days	Unlimited	<1 min
Model D	\$0–\$50 (?)	2 h/ 90 days	2 · 10 (?)	0

Rather than discussion via forum on the conference Web page, Model B would involve only presentation on the Web site, and discussion would occur via e-mail.

The third model for the virtual conference, Model C, accelerates the discussion even more. In this model, discussion would be handled via a "Chat server," which would be required in addition to the Web server used in Models A and B. The Chat server would moderate live text "chat" among the participants on a real-time basis. The "chat room" would be structured so that all interested participants would discuss the presentation during a specific time period.

Finally, yet another model for the virtual conference, Model D, is that of a live video conference, where both presentation and discussion would be held via streaming video/audio. We envision that this would most likely occur through the distance learning facilities that are now widely available on college campuses in the US. In this model, the presenter would give a live video presentation from the transmitting facility, to both a live audience and to recipients of the broadcast at other facilities. The video/audio would be streamed to each facility, and the participants could then participate in discussion on a real-time basis.

To compare the various conference models, we have created Table 2. Model N (N stands for Nassau) represents a live conference. The traditional conference usually has a substantial price tag for registration, travel, meals, etc. The duration of a live conference is limited, for one can only attend so many sessions per day; we are also limited in number by the physical constraints of the conference site. The major benefit of a live conference is apparent – face-to-face meetings and real-time discussion with colleagues can be invaluable. However, when comparing a live conference with the various virtual conference models (Models A–D), we see that none of the disadvantages of a live meeting would be present: the price tag for participants would be substantially lower for a virtual conference, and with a computer-based virtual conference model the conference could be held at any time, for any duration, and with any number of participants. However, there is usually some lag time involved in the discussion mode (up to a few hours), in order to allow for server processing time, and the face-to-face meetings simply do not occur.

Regional Web Page Moving on to some of the other ideas to build community, the idea of a Regional Web Page is appealing from an informational perspective. The US Regional Web Page could be hosted from the Web site recently developed by MEDC and IBAME http://www.mossbauer.org. It is fairly simply to maintain and will cost nothing except for the time it takes to assemble and upload the data. In this way, the US community could keep abreast of each other's research and other news that could be posted on the site from time to time, such as current research interests, job openings, visiting scientists, requests for information about sources and instrumentation, etc.

Note that MEDC has planned for many Regional Pages on the http://www.mossbauer.org site. For example, the Regional Page for Germany currently contains links to various individual German lab Web sites. The US Regional Page would exist on the site along with the other Regional Pages and could also contain links to individual lab Web sites, as well as the other information mentioned previously.

In order to implement the US Regional Web Page, we would need a volunteer to act as Regional Editor to assemble the information and provide it to MEDC. Our staff has volunteered its time and resources to uploading and maintaining the site from the MEDC server. Currently, several Mössbauer researchers have volunteered to act as Regional Editors for their respective geographic regions. They are: Stewart Campbell from Australia, Denis Rancourt from Canada, Peter Schaaf from Germany, and George Fern from the UK.

Newsletter Another idea, similar to that of the Regional Web Page, would be to publish, either electronically or via hard copy, a US Mossbauer Newsletter that would include news and information of interest to the US Mössbauer community. The Newsletter could contain the same type of information as discussed above for the US Regional Web Page, such as current research interests, position openings, visiting scientists, questions regarding the availability or repair of instrumentation and sources, and general questions to be posed to the US Mössbauer community. Such a Newsletter could be published as needed, but we believe it would be most effective if it were published either quarterly or semi-annually.

Like the Regional Web Page, in order to implement a US Mössbauer Newsletter we would need to solicit someone to act as an Editor to solicit and compile information. The MEDC staff has volunteered to handle the layout, editing, proofreading, and distribution.

4 Conclusion

To conclude, the US remains as one of the strongest contributors to Mössbauer research on the global level. However, there are some alarming trends developing. As we have seen, the number of active Mossbauer researchers has begun a decline, and we also see the relative isolation of many US Mossbauer groups. The attendance of US researchers at the ICAME conference series is an indication of this.

New solutions are needed to overcome the isolation of US Mössbauer researchers as we move forward to strengthen the community. We have discussed some possible solutions, including the continuation of the Nassau conference series, along with the implementation of some new ideas, such as virtual conferences or the publication of a US Regional Web Page or Newsletter.

References

1. Frauenfelder, H.: Mössbauer effect: recoilless emission and absorption of gamma rays. In: Frauenfelder, H., Lustig, H. (eds.) Proceedings of the First International Conference on the Mössbauer Effect, Monticello, Illinois, USA, p. iii. University of Illinois (1960)
2. Taylor, R.D.: MERDJ. **26**(7), 213 (2003)

Hyperfine Interact (2006) 170:169–177
DOI 10.1007/s10751-007-9508-5

Two earth years of Mössbauer studies of the surface of Mars with MIMOS II

G. Klingelhöfer · R. V. Morris · P. A. De Souza Jr. ·
D. Rodionov · C. Schröder

Published online: 31 January 2007

Abstract The element iron plays a crucial role in the study of the evolution of matter from an interstellar cloud to the formation and evolution of the planets. In the Solar System iron is the most abundant metallic element. It occurs in at least three different oxidation states: Fe(0) (metallic iron), Fe(II) and Fe(III). Fe(IV) and Fe(VI) compounds are well known on Earth, and there is a possibility for their occurrence on Mars. In January 2004 the USA space agency NASA landed two rovers on the surface of Mars, both carrying the Mainz Mössbauer spectrometer MIMOS II. They performed for the first time in-situ measurements of the mineralogy of the Martian surface, at two different places on Mars, Meridiani Planum and Gusev crater, respectively, the landing sites of the Mars-Exploration-Rovers (MER) Opportunity and Spirit. After about two Earth years or one Martian year of operation the Mössbauer (MB) spectrometers on both rovers have acquired data from more than 150 targets (and more than thousand MB spectra) at each landing site. The scientific measurement objectives of the Mössbauer investigation are to obtain for rock, soil, and dust (1) the mineralogical identification of iron-bearing phases (e.g., oxides, silicates, sulfides, sulfates, and carbonates), (2) the quantitative measurement of the distribution of iron among these iron-bearing phases (e.g., the relative proportions of iron in olivine, pyroxenes, ilmenite and magnetite in a basalt), (3) the quantitative measurement of the distribution of iron among its oxidation states (e.g., Fe^{2+}, Fe^{3+}, and Fe^{6+}), and (4) the characterization of the size distribution of magnetic particles. Special geologic targets of the Mössbauer investigation are dust collected by the Athena magnets and interior rock and soil surfaces exposed by the Athena Rock Abrasion Tool and by trenching with rover wheels. The Mössbauer spectrometer on *Opportunity* at Meridiani Planum, identified eight Fe-bearing phases: jarosite

G. Klingelhöfer (✉) · D. Rodionov · C. Schröder
Institut Inorganic and Analytical Chemistry, Johannes Gutenberg-University Mainz, Staudinger Weg 9, 55099 Mainz, Germany
e-mail: klingel@mail.uni-mainz.de

R. V. Morris
NASA Johnson Space Center, Houston, TX, USA

P. A. De Souza Jr.
CVRD, Rio de Janeiro, RJ, Brazil

(K,Na,H3O)(Fe,Al)(OH)6(SO4)2, hematite, olivine, pyroxene, magnetite, nanophase ferric oxides (npOx), an unassigned ferric phase, and a metallic Fe–Ni alloy (kamacite) in a Fe–Ni-meteorite. Outcrop rocks consist of hematite-rich spherules dispersed throughout S-rich rock that has nearly constant proportions of Fe^{3+} from jarosite, hematite, and npOx (28%, 35%, and 19% of total Fe). Jarosite is mineralogical evidence for aqueous processes under acid–sulfate conditions because it has structural hydroxide and sulfate and it forms at low pH. Hematite-rich spherules, eroded from the outcrop, and their fragments are concentrated as hematite-rich soils (lag deposits) on ripple crests (up to 68% of total Fe from hematite). Olivine, pyroxene, and magnetite are primarily associated with basaltic soils and are present as thin and locally discontinuous cover over outcrop rocks, commonly forming aeolian bed-forms. Basaltic soils are more reduced (Fe^{3+}/Fe_{total} ~0.2–0.4), with the fine-grained and bright aeolian deposits being the most oxidized. Basaltic soil at Meridiani Planum and Gusev crater have similar Fe-mineralogical compositions. At Gusev crater, the Mössbauer spectrometer on the MER Spirit rover has identified 8 Fe-bearing phases. Two are Fe^{2+} silicates (olivine and pyroxene), one is a Fe^{2+} oxide (ilmenite), one is a mixed Fe^{2+} and Fe^{3+} oxide (magnetite), two are Fe^{3+} oxides (hematite and goethite), one is a Fe^{3+} sulfate (mineralogically not constrained), and one is a Fe^{3+} alteration product (npOx). The surface material in the plains have a olivine basaltic signature (Morris, et al., *Science,* 305: 833, 2004; Morris, et al., *J. Geophys. Res.,* 111, 2006, Ming, et al., *J. Geophys. Res.*, 111, 2006) suggesting physical rather than chemical weathering processes present in the plains. The Mössbauer signature for the Columbia Hills surface material is very different ranging from nearly unaltered material to highly altered material. Some of the rocks, in particular a rock named Clovis, contain a significant amount of the Fe oxyhydroxide goethite, α-FeOOH, which is mineralogical evidence for aqueous processes because it is formed only under aqueous conditions.

Key words MIMOS II · weathering · Mars · jarosite · goethite · hematite · field distribution · Gusev crater · Meridiani Planum · Mars-exploration-rovers · mineralogy · instrumentation · backscattering

1 Introduction

The element Fe is one of the most abundant elements in the universe. Iron is the sixth most abundant element in the Earth's crust and the third most abundant cationic element after Si and Al. There is, therefore, hardly any rock completely free from Fe [10]. Other than Earth, the Red Planet, in particular, owes its colour to Fe-oxides. Mars surface materials are enriched in Fe relative to Earth (e.g. [11–16]). Fe Mössbauer spectroscopy determines the oxidation state of the element Fe, identifies Fe-bearing mineral phases, measures the relative abundance of iron among those phases, and yields some information on particle size and crystalinity, thus providing a useful tool for the study of weathering phenomena on Earth and Mars [17]. The distribution of Fe between Fe-bearing minerals and its oxidation states constrains the primary rock type (e.g., olivine-bearing versus non-olivine-bearing basalt), the redox conditions under which primary minerals crystallized (e.g., presence or absence of magnetite), the extent of alteration and weathering (e.g., value of Fe^{3+}/Fe_{total}), the type of alteration and weathering products (e.g., oxides versus sulfates versus phyllosilicates), and the processes and environmental conditions for alteration and weathering (e.g., neutral versus acid-chloride versus acid–sulfate aqueous process under ambient or hydrothermal conditions) [18].

In January 2004 the USA space agency NASA landed successfully two rovers on the surface of Mars. The Mars-Exploration-Rovers (MER) Spirit at the Gusev crater landing site and Opportunity at the Meridiani Planum landing site are both carrying our Mössbauer spectrometer MIMOS II mounted on an robotic arm (IDD) on the rovers [1, 8]. The primary objective of the Athena science investigation is to explore two sites on the Martian surface where water may once have been present, and to assess past environmental conditions at those sites and their suitability for life [8].

The IDD instruments, in particular the Mössbauer spectrometer and the APXS (X-ray fluorescence for elemental analysis [8, 14–16]) are used to determine the chemistry and mineralogy of rocks and soils. Both the rovers and instruments have been designed to operate for about 90 sols (1 sol=1 Martian day) but were still completely functional and operating in spring 2006 (more than 2 Earth years, or one Martian year, of continues operation) and are expected to last for several more month. The performance of the Mössbauer spectrometers MIMOS II did not degrade, only the intensity if the Co57/Rh source, with an activity of about 340 mCi at launch in June 2003, decreased according to the half life.

2 The MIMOS II spectrometer

The MER MIMOS II Mössbauer spectrometers (see Fig. 1) have been described in more detail in [1, 19]. Briefly, Opportunity's and Spirit's spectrometers operate in backscatter geometry (Fig. 2). Because of the complexity of sample preparation, backscatter measurement geometry is the choice for an in situ planetary Mössbauer instrument (e.g. [7, 19]. No sample preparation is required, because the instrument is simply presented to the sample for analysis. Data are acquired within 512 velocity channels, using a triangular waveform at a drive frequency of ~24 Hz, and a standard but selectable velocity range of about ±12 mm/s (slightly different for Spirit and Opportunity, respectively), and binning of data into 13 temperature windows with 10 K width, to account for the temperature dependence of the MB hyperfine parameters. The instruments were equipped with primary ^{57}Co radiation sources with ~330 mCi at launch (~30 mCi on sol 557), and with four detectors to detect resonantly scattered X-rays (6.4 keV) and γ-rays (14.41 keV) from surface targets. Both 6.4 and 14.41 keV data were returned to the Earth because the detector counters were configured to sum, pairwise, counts from the 6.4 and 14.41 keV energy windows.

3 Meridiani Planum landing site

The surface of Meridiani Planum explored by the Opportunity rover can be described as a flat plain of S-rich outcrop that is mostly covered by thin surficial deposits of aeolian basaltic sand and dust, and lag deposits of hard Fe-rich spherules (and fragments thereof) that weathered from the outcrop and small unidentified rock fragments and cobbles (e.g. [20]. Surface expressions of the outcrop occur at impact craters (e.g., Eagle, Fram, Endurance, Erebus craters) and occasionally in troughs between ripple crests.

Mössbauer measurements show that the Fe-containing material in the outcrop consists predominantly of the Fe-sulfate jarosite, hematite, and a basaltic component (olivine, pyroxene) (Fig. 3a,b). The same material was found again along the several kilometers long drive way of Opportunity, in particular at craters Fram and Endurance, suggesting that the whole area is covered with this jarositic material. The mineral jarosite ($(K,Na)Fe_3(SO_4)_2(OH)_6$)

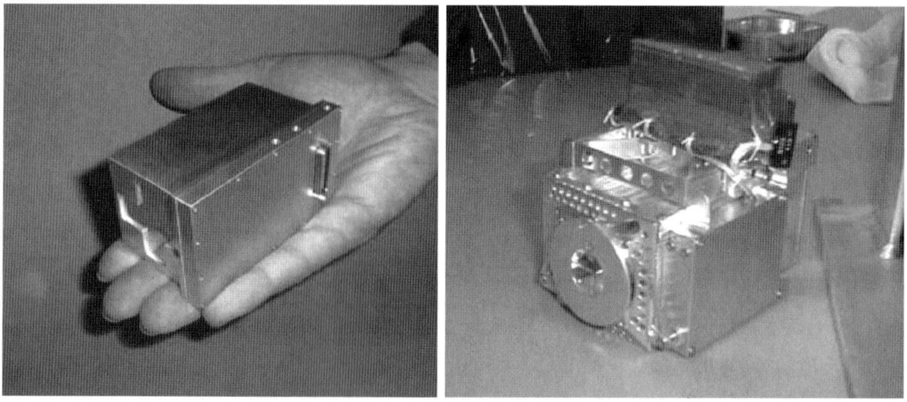

Fig. 1 The miniaturized Mössbauer spectrometer MIMOS II sensor head: (*left*) the sensor head itself on real human hand; (*right*) MIMOS II flight unit 3 (the Opportunity instrument) with the contact plate at the front. The contact plate is integrated to a contact switch arrangement (*two boxes at the left and right side* of the instrument body (*only the box on the right side* can be seen)). The inner opening of the contact plate determines the area of the sample to be analyzed and has a diameter of 15 mm

contains hydroxyl and is thus direct mineralogical evidence for aqueous, acid–sulfate alteration process on early Mars.

The plains and a large portion of Eagle and Endurance crater are covered by spherules, nicknamed Blueberries, with a diameter of several mm. Mössbauer data clearly show that the composition of these spherules is dominated by the Fe-oxide hematite [3, 4]. The composition of the soil at Meridiani is found to be basaltic, dominated by olivine similar to the Gusev site, but in some areas the soil is enriched in the mineral hematite (hematitic soil). The reason for this is not clear yet. The Fe^{2+}/Fe_{total} ratio at Meridiani Planum varies from ~0.8 for the basaltic soil to ~0.1 for the outcrop.

In Fig. 3 we show representative MB spectra for Meridian Planum along with the component subspectra obtained from the fitting procedures. Generic names for the subspectra are used with the format FeXYZ, where X=2, 2.5, or 3 for the Fe oxidation state, Y = D or S (doublet or sextet), and Z is a sequential number for subspectra having the same FeXY (for details see [18, 21]. We were able to describe all Meridiani Planum spectra with five doublet and four sextet subspectra: two doublets from octahedrally coordinated Fe^{2+} (oct-Fe^{2+}) (Fe2D1 and Fe2D2), three doublets from oct-Fe^{3+} (Fe3D1, Fe3D3, and Fe3D4), one sextet from tetrahedrally coordinated Fe^{3+} (tet-Fe^{3+}) (Fe3S1), one sextet from coupled oct-Fe^{2+} and Fe^{3+} (Fe2.5S1), one sextet from oct-Fe^{3+} (Fe3S2), and one sextet from Fe–Ni metal (Fe0S1). This description assumes a one-doublet fit (using Fe2D2) for spectra from Bounce Rock. A superior fit is obtained with a two-doublet model (both oct-Fe^{2+}). The fitting of individual Meridiani Planum spectra required between two and seven subspectra. Doublets Fe3D3 and Fe3D4 and sextet Fe0S1 were not observed at Gusev crater.

4 Gusev crater landing site

The rover Spirit landed on the floor of Gusev Crater, a large crater with about 160 km diameter. Ma'adim Vallis, a large valley network, is cutting through the southern rim of Gusev. It is believed that once water flew through this valley network into Gusev crater and sediments carried by the water are expected to have settled on the crater floor. Therefore

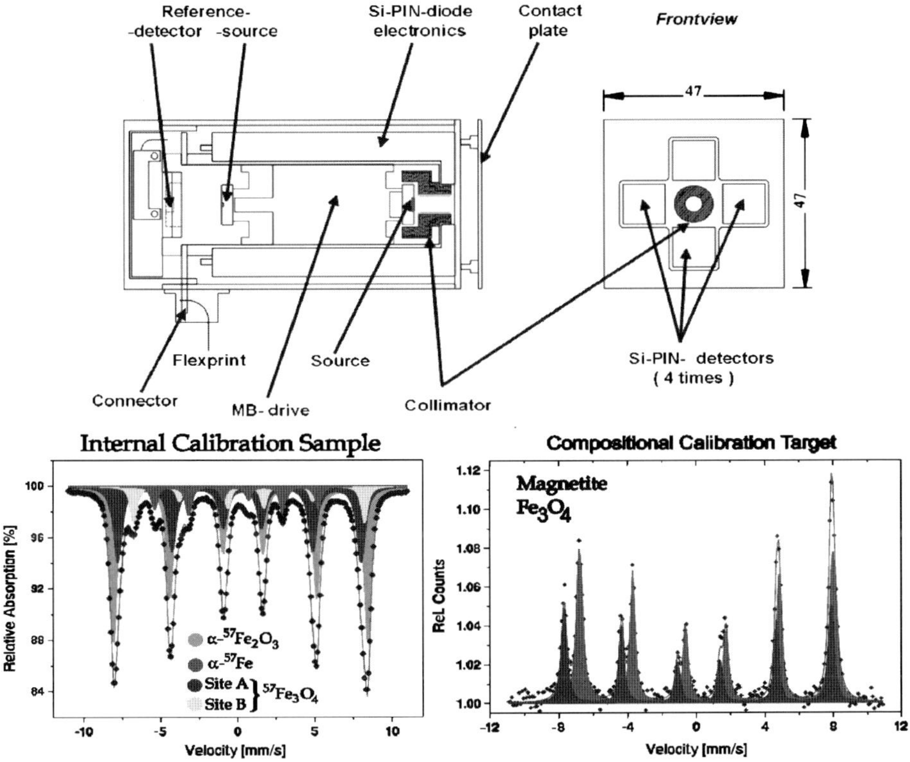

Fig. 2 Scheme of the miniaturized Mössbauer spectrometer MIMOS II; a typical spectrum of the internalCalibration channel is shown, as well as a backscatter spectrum taken on the CCT (compositional calibration target) mounted below the rover deck. The CCT is completely composed of magnetite (Fe_3O_4)

exploration of the crater floor was expected to offer opportunities to find and study sediments and other material altered by aqueous processes.

During the first Martian year (~700 sols, 1 sol–1 martian day, two Earth years.–~1 Martian year) of operation at the Gusev crater landing site the *Spirit* Mössbauer instrument analysed rocks and soils on the crater floor, at a crater close to the landing site called Bonneville Crater, the ejecta plains between Bonneville and the Columbia Hills before entering the hills. After about 160 sols Spirit reached the Columbia Hills and started climbing up to the summit, traversing across the saddle to Husband Hill and up to the rocks and soils on the Cumberland ridge on the northwest side of Husband Hill. From there Spirit descended the Columbia Hills on the other side on its way to a large dune field called El-Dorado and Home Plate, a light-toned plateau ~90 m in diameter and 2–3 m high within the Inner Basin of the Columbia Hills. During this more than 500 sols traverse through Husband Hill MIMOS II acquired data from a large number of soils and rocks. Representative MB spectra are shown in Fig. 4.

As determined by MB already in the very first spectrum and confirmed by many others [2, 4], the rocks and soil around Bonneville and in the cratered plains have a basaltic MB signature, and their Mössbauer spectra are dominated by an olivine doublet [2, 4, 5]. The presence of abundant olivine in rocks and in surrounding soil as determined by MB suggests that physical rather than chemical weathering processes currently dominate the plains at Gusev crater.

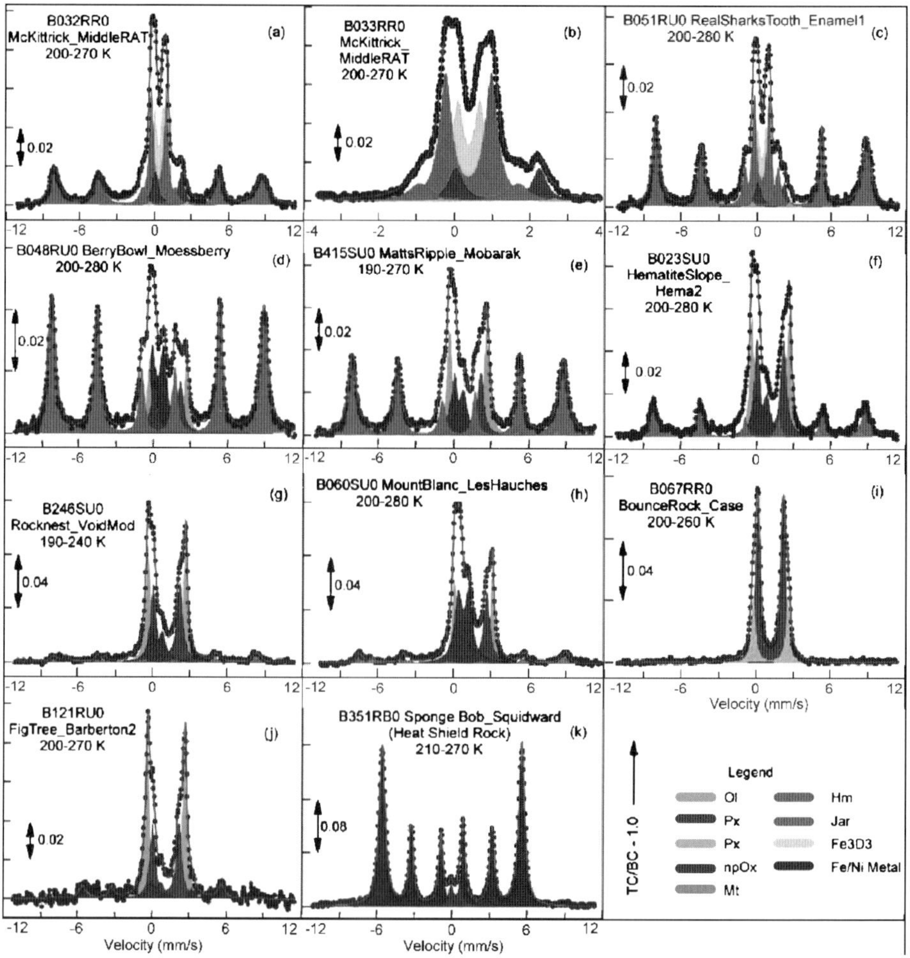

Fig. 3 Meridiani Planum Mössbauer mineralogy. The large variability in mineral composition at this landing site can be seen. Shown are representative Mössbauer spectra. Spectra are the sum over all temperature windows within the indicated temperature ranges. The computed fit and component subspectra (Lorentzian lineshapes) from least-squares analyses are shown by the *solid line* and the *solid shapes*, respectively. Full (**a**) and reduced (**b**) velocity MB spectra for interior Burns outcrop exposed by RAT grinding show that hematite, jarosite, and Fe3D3 (unidentified Fe^{3+} phase) are the major Fe-bearing phases. MB spectrum for a rind or crust (**c**) of outcrop material that has an increased contribution of hematite relative to jarosite plus Fe3D3. MB spectra of two Moessberry Subclass soils (**d** and **e**) have high concentrations of hematite. The spectrum (**e**) is typical for hematite lag deposits at ripple crests. The spectrum of Nougat Subclass soil (**f**) is transitional between Moessberry Subclass soil and Laguna Class soil. Spectra of the Panda (**g**) and Gobi (**h**) Subclasses of Laguna Class soil are basaltic in nature and have olivine, pyroxene, and npOx as major Fe-bearing phases. MountBlanc_LesHauches is considered to be enriched in martian dust. MB spectra in (**i**–**k**) are for three single-occurrence rocks. BounceRock (**i**) is monomineralic pyroxene with respect to Fe-bearing phases. Barberton (**j**) contains kamacite (iron–nickel metal), and Heat Shield Rock is nearly monomineralic kamacite (Fe–Ni metal alloy with ~5% Ni), probably a Fe–Ni meteorite. TC=total counts and BC=baseline counts (from [21])

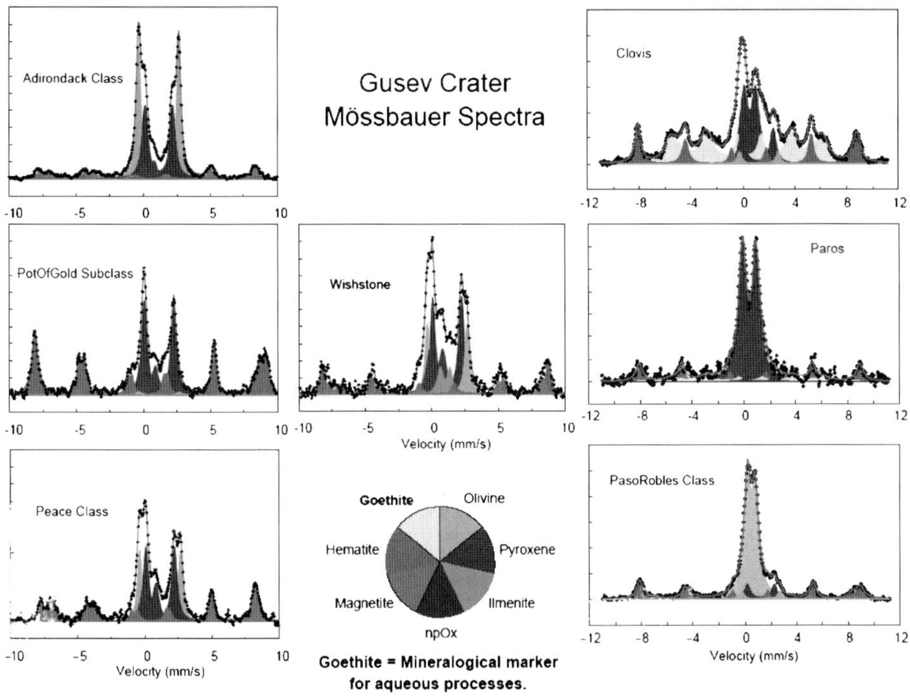

Fig. 4 Typical Gusec Crater Mössbauer spectra. Clearly the great variability in mineral composition can be seen [2, 5]. On the *upper left* the spectrum of the Adirondack class material (olivine basaltic composition), which dominates the soil and rocks in the plain, is shown. All other examples of spectra are from the Columbia Hills area. These materials are highly weathered. The spectrum on the lower right, taken from a soil called Paso Robles is dominated by a Fe–Sulfate doublet

The MB signatures for the Columbia Hills are very different. Instead of relatively fresh olivine basalt, the MB signature is characteristic of highly altered rocks. In particular, one of the rocks, named Clovis, which was found in the West Spur region, contains the highest amount of the Fe oxyhydroxide goethite (Gt), α-FeOOH, of about 40% in area (see Fig. 4) found to date in the Columbia Hills. Detailed analyses of the data ([9] and references therein) using superpositions of two model-independent hyperfine-field distributions (HFD) with two ditto quadrupole-splitting distributions (QSD), have indeed yielded hyperfine parameters that are in line with those of synthetic α-FeOOH having a particle size of ~10 nm. Like jarosite found at Meridiani Planum [3], the mineral goethite is a marker for aqueous processes as goethite forms only in the presence of water, and it contains the hydroxide anion. The rock Clovis also contains a significant amount of hematite [9].

During the first Martian year (~2 Earth years) at Gusev Crater the Mössbauer spectrometer on the MER Spirit rover has identified 8 Fe-bearing phases (see Fig. 4). Two are Fe^{2+} silicates (olivine and pyroxene), one is a Fe^{2+} oxide (ilmenite), one is a mixed Fe^{2+} and Fe^{3+} oxide (magnetite), two are Fe^{3+} oxides (hematite and goethite), one is a Fe^{3+} sulfate (mineralogically not constrained), and one is a Fe^{3+} alteration product (npOx). We envision that NpOx is poorly crystalline (amorphous) without distinct mineralogical composition and with a diffuse chemical composition that depends on local conditions. Element associations indicate npOx has affinity for sulfate and chloride anion either through the formation of chemical bonds or specific chemical adsorption. The behavior of hematite is complex because the temperature of the Morin transition (~260 K) lies within diurnal temperature variations on Mars. Goethite is the only unequivocal H_2O/OH bearing phase, and it yields ~11% H_2O upon dehydroxalation. As an alteration product, npOx also likely contains H_2O/OH, but the evidence is inferential [5, 6]. A hydrous Fe^{3+}-sulfate is inferred form combined MB, APXS, Mini-TES, and Pancam data.

5 Summary

The nominal mission for both Spirit and Opportunity was scheduled for 90 Mars days or sols and a traverse of about 600 m. Both rover missions have now exceeded significantly the nominal mission by operating more than 700 sols on Mars. Spirit travelled a distance of about 5 km, and Opportunity more than 6 km.

Through sol 700 the MER Spirit Mössbauer spectrometer MIMOS II has identified different Fe-Oxides, silicates and other minerals and Fe-compounds, in total 8 different phases, up to now at the Gusev landing site. Two are Fe^{2+} silicates (olivine and pyroxene), one is a Fe^{2+} oxide (ilmenite), one is a mixed Fe^{2+} and Fe^{3+} oxide (magnetite), two are Fe^{3+} oxides (hematite and goethite), one is a Fe^{3+} sulfate (mineralogically not constrained), and one is a Fe^{3+} alteration product (npOx). The behavior of hematite is complex because the temperature of the Morin transition (~260 K) lies within diurnal temperature variations on Mars. The spacial distribution of the minerals show that the plains of Gusev crater are mainly basaltic and relatively unaltered with an $Fe^{3+}/Fe_{total}<0.2$. The Columbia hills on the other hand show a high degree of alteration. Only a few rocks are relatively unaltered. One of the highest degrees of alteration has been seen at the Clovis outcrop with a $Fe^{3+}/Fe_{total}\sim 0.8$. Another striking feature of Columbia Hill's outcrop is a variability of Fe-oxide phases for different locations. For example, Clovis has no magnetite component. Ebenezer, a different rock target in the hills, however, has substantial magnetite contribution in addition to hematite and goethite. The rock Clovis contains a significant amount of goethite with about 40% of the total spectral area (see Fig. 2). The mineral goethite is mineralogical evidence for the presence of H_2O/OH on Mars and for aqueous processes in the history of Mars.

The MB instrument on Opportunity has analyzed more than 100 rock an soil targets at Meridiani Planum during the first Martian year of operation. Eight Fe-bearing phases were identified: kamacite as a Fe^0-bearing phase in a Fe–Ni meteorite, olivine and pyroxene as Fe^{2+}-bearing phases, magnetite as a Fe^{2+}- and Fe^{3+}-bearing phase, and nanophase oxide (npOx), hematite, jarosite, and Fe3D3 as Fe^{3+}-bearing phases. The oxidation state of Fe ranged from $Fe^{3+}/FeT=0.00$ to 0.93. The least oxidized targets ($Fe^{3+}/FeT\leq0.06$) are isolated rocks (BounceRock, Barberton, and Heat Shield rock). Surfaces of S-rich outcrop rocks exposed by RAT grinding are very oxidized with average $Fe^{3+}/FeT = 0.85 \pm 0.03$. The presence of jarosite is mineralogical evidence for the presence of H2O/OH on Mars and for acid–sulfate aqueous processes.

Both rovers and the instruments MIMOS II continue the exploration of the two landing sites and are expected to pass the 1000 sol limit after surviving the second Martian winter.

Acknowledgments This work was funded by the German Space Agency DLR under contract 50QM99022. The support of the TU Darmstadt, Germany, and the University of Mainz, Germany is acknowledged. R.V.M. acknowledges the support of the NASA Mars Exploration Rover Project and the NASA Johnson Space Center.

References

1. Klingelhöfer, G., Morris, R.V., Bernhardt, B., Rodionov, D., De Souza, P.A. Jr., Squyres, S.W., Foh, J., Kankeleit, E., Bonnes, U., Gellert, R., Schröder, C., Linkin, S., Evlanov, E., Zubkov, B., Prilutski, O.: J. Geophys. Res. **108**, 8067 (2003)
2. Morris, R.V., Klingelhöfer, G., Bernhardt, B., Schröder, C., Rodionov, D.S., De Souza, P.A. Jr., Yen, A., Gellert, R., Evlanov, E.N., Foh, J., Kankleit, E., Gütlich, P., Ming, D.W., Renz, F., Wdowiak, T., Squyres, S.W., Arvidson, R.E.: Science **305**, 833 (2004)
3. Klingelhöfer, G., Morris, R.V., Bernhardt, B., Schröder, C., Rodionov, D.S., De Souza, P.A. Jr., Yen, A., Gellert, R., Evlanov, E.N., Zubkov, B., Foh, J., Bonnes, U., Kankleit, E., Gütlich, P., Ming, D.W., Renz, F., Wdowiak, T., Squyres, S.W., Arvidson, R.E.: Science **306**, 1740 (2004)
4. Klingelhöfer, G.: In: Garcia, M., Marco, J.F., Plazaola, F. (eds.) Industrial Applications of the Mössbauer Effect. American Institute of Physics (2005)
5. Morris, R.V., Klingelhöfer, G., Schröder, C., Rodionov, D.S., Yen, A., Ming, D.W., De Souza, P.A. Jr., Fleischer, I., Wdowiak, T., Gellert, R., Bernhardt, B., Evlanov, E.N., Zubkov, B., Foh, J., Bonnes, U., Kankeleit, E., Gütlich, P., Renz, F., Squyres, S.W., Arvidson, R.E.: J. Geophys. Res. **111**, (2006) E02S12, doi:10.1029/2005JE002560
6. Ming, D.W., Mittlefehldt, D.W., Morris, R.V., Golden, D.C., Gellert, R., Yen, A., Clark, B.C., Squyres, S.W., Farrand, W.H., Ruff, S.W., Arvidson, R.E., Klingelhöfer, G., McSween, H.Y., Rodionov, D.S., Schröder, C., De Souza, P.A. Jr., Wang, A.: J. Geophys. Res. **111**, (2006) E02S12, doi:10.1029/2005JE002560
7. Klingelhöfer, G., Held, P., Teucher, R., Schlichting, F., Foh, J., Kankeleit, E.: Hyperfine Interact. **95**, 305–339 (1995)
8. Squyres, S.W., et al.: J. Geophys. Res. **108**, 8062 (2003) doi:10.1029/2003JE002121
9. Klingelhöfer, G., DeGrave, E., Morris, R.V., Van Alboom, A., de Resende, V.A., De Souza, P.A. Jr., Rodionov, D., Schröder, C., Ming, D.W., Yen, A., Hyperfine Interact. (2006) doi:10.1007/s10751-006-9329-y
10. Cornell, R.M., Schwertmann, U.: The Iron Oxides. VCH Verlagsgesellschaft mbH, Weinheim, Germany (1996)
11. Clark, B.C., Baird, A.K., Rose, H.J., Toulmin, P. III, Keil, K., Castro, A.J., Kelliher, W.C., Rowe, C.D., Evans, P.H.: Science **194**, 1283–1288 (1976)
12. Toulmin, P. III, Baird, A.K., Clark, B.C., Keil, K., Rose, H.J. Jr., Christian, R.P., Evans, P.H., Kelliher, W.C.: J. Geophys. Res. **84**, 4625–4634 (1977)
13. Rieder, R., Economou, T., Wänke, H., Turkevich, A., Crisp, J., Brückner, J., Dreibus, G., McSween, H. Y. Jr.: Science **278**, 1771–1774 (1997)
14. Rieder, R., Gellert, R., Brückner, J., Klingelhöfer, G., Dreibus, G., Yen, A., Squyres, S.W.: J. Geophys. Res. **108**(E12), 8066 (2003) doi:10.1029/2003JE002150
15. Gellert, R., Rieder, R., Anderson, R.C., Brückner, J., Clark, B.C., Dreibus, G., Economou, T., Klingelhöfer, G., Lugmair, G.W., Ming, D.W., Squyres, S.W., d'Uston, C., Wänke, H., Yen, A., Zipfel, J.: Science **305**, 829–832 (2004)
16. Gellert, R., Rieder, R., Brückner, J., Clark, B.C., Dreibus, G., Klingelhöfer, G., Lugmair, G.W., Ming, D. W., Wänke, H., Yen, A., Zipfel, J., Squyres, S.W.: J. Geophys. Res. **111**, (2006) E02S05, doi:10.1029/2005JE002555
17. Schröder, C., Klingelhöfer, G., Tremel, W.: Planet. Space Sci. **52**(11), 997–1010 (2004) doi:10.016/j.pss.2004.07.018
18. Morris, R.V., Klingelhöfer, G., Schröder, C., Rodionov, D.S., Yen, A., Ming, D.W., De Souza, P.A. Jr., Fleischer, I., Wdowiak, T., Gellert, R., Bernhardt, B., Evlanov, E.N., Zubkov, B., Foh, J., Bonnes, U., Kankeleit, E., Gütlich, P., Renz, F., Squyres, S.W., Arvidson, R.E.: J. Geophys. Res. **111**, (2006) E02S13, doi:10.1029/2005JE002584
19. Klingelhöfer, G., Fegley, B. Jr., Morris, R.V., Kankeleit, E., Held, P., Evlanov, E., Priloutskii, O.: Planet. Space Sci. **44**, 1277–1288 (1996)
20. Squyres, S.W., Grotzinger, J.P., Arvidson, R.E., Bell, J.F. III, Calvin, W., Christensen, P.R., Clark, B.C., Crisp, J.A., Farrand, W.H., Herkenhoff, K.E., Johnson, J.R., Klingelhöfer, G., Knoll, A.H., McLennan, S.M., McSween, H.Y., Morris, R.V., Rice, J.W., Rieder, R., Soderblom, L.A.: Science **306**, 1709–1714 (2004)
21. Morris, R.V., Klingelhöfer, G., Schröder, C., Rodionov, D.S., Yen, A., Ming, D.W., De Souza, P.A. Jr., Wdowiak, T., Fleischer, I., Gellert, R., Bernhardt, B., Bonnes, U., Cohen, B.A., Evlanov, E.N., Foh, J., Gütlich, P., Kankeleit, E., McCoy, T., Mittlefehldt, D.W., Renz, F., Schmidt, M.E., Zubkov, B., Squyres, S.W., Arvidson, R.E.: J. Geophys. Res. **112**, (2007) doi:10.1029/2006JE002791

Author Index to Volume 170 (2006)